algebra for all purple level

elizabeth warren PhD

About the Author

Elizabeth Warren has been involved in Mathematics Education for more than 30 years. During this time she has actively worked in both the university and school levels and engaged both elementary and secondary schoolteachers in professional development activities. She is presently conducting research into Patterns and Algebra.

Algebra for All, *Purple Level*

Author: Elizabeth Warren, PhD

Warren, Elizabeth.
Algebra for All: purple level.
ISBN 1 921023 03 1.
1. Algebra - Problems, exercises, etc. - Juvenile literature. I. Title.
512

For more information, contact
North America
Tel. 1-888-ORIGO-01 or 1-888-674-4601
Fax 1-888-674-4604
sales@origomath.com
www.origomath.com

Australasia
For more information,
email info@origo.com.au
or visit www.origo.com.au for other contact details.

ISBN: 1 921023 03 1

10 9 8 7 6 5 4 3 2 1

INTRODUCTION

What is algebra?

Algebraic thinking commences as soon as students identify consistent change and begin to make generalizations. Their first generalizations relate to real-world experiences. For example, a child may notice a relationship between her age and the age of her older brother. In the example below, Ali has noted that her brother Brent is always 2 years older than her.

Ali's age	Brent's age
8	10
9	11
10	12
11	13

Over time these generalizations extend to more abstract situations involving symbolic notation that includes numbers. The above relationship can be generalized using the following symbolic notation.

Ali + 2 = Brent ***A + 2 = B***

Algebraic thinking uses different symbolic representations, such as unknowns and variables, with numbers to explore, model, and solve problems that relate to change and describe generalizations. The symbol system used to describe generalizations is formally known as algebra. Following the example above, Ali wonders how old she will be when Brent is 21 years old. We can solve a problem such as this by "backtracking" the generalization (A = 21 – 2).

Algebra involves the generalizations that are made regarding the relationships between variables in the symbol system of mathematics.

Why algebra?

Identifying patterns and making generalizations are fundamental to all mathematics, so it is essential that students engage in activities involving algebra. Many practical uses for algebra lie hidden under the surface of an increasingly electronic world, such as specific rules used to determine telephone charges, track bank accounts and generate statements, describe data represented in graphs, and encrypt messages to make the Internet secure. Algebraic thinking is more overt when we create rules for spreadsheets or simply use addition to solve a subtraction problem.

What are the "big ideas"?

The lessons in the *Algebra for All* series aim to develop the "big ideas" of early algebra while supporting thinking, reasoning, and working mathematically. These ideas of equivalence and equations, patterns and functions, properties, and representations are inherent in all modern curricula and are summarized in the following paragraphs.

Equivalence and Equations

The most important ideas about equivalence and equations that students need to understand are:

- "Equals" indicates equivalent sets rather than a place to write an answer
- Simple real-world problems with unknowns can be represented as equations
- Equations remain true if a consistent change occurs to each side (the balance strategy)
- Unknowns can be found by using the balance strategy.

Patterns and Functions

This idea focuses on mathematics as "change". Change occurs when one or more operation is used. For example, the price of an item bought on the Internet changes when a freight charge is added. It is important for students to understand that:

- Operations almost always change an original number to a new number
- Simple real-world problems with variables can be represented as "change situations"
- "Backtracking" reverses a change and can be used to solve unknowns.

Properties

Students will discover a variety of arithmetic properties as they explore number, such as:

- The commutative law and the associative law exist for addition and multiplication but not for subtraction and division
- Addition and subtraction are inverse operations, as are multiplication and division
- Adding or subtracting zero and multiplying or dividing by 1 leaves the original number unchanged
- In certain circumstances, multiplication and division distribute over addition and subtraction.

Representations

Different representations deepen our understanding of real-world problems and help us identify trends and find solutions. This idea focuses on creating and interpreting a variety of representations to solve real-world problems. The main representations that are developed in this series include graphs, tables of values, drawings, equations, and everyday language.

INTRODUCTION

About the series

Each of the six *Algebra for All* books features 4 chapters that focus separately on the "big ideas" of early algebra — Equivalence and Equations, Patterns and Functions, Properties, and Representations. Each chapter provides a carefully structured sequence of lessons. This sequence extends across the series so that students have the opportunity to develop their understanding of algebra over a number of years.

About the lessons

Each lesson is described over 2 pages. The left-hand page describes the lesson itself, including the aim of the lesson, materials that are required, clear step-by-step instructions, and a reflection. These notes also provide specific questions that teachers can ask students, and subsequent examples of student responses. The right-hand page supplies a reproducible blackline master to accompany the lesson. The answers for all blackline masters can be found on pages 66-73.

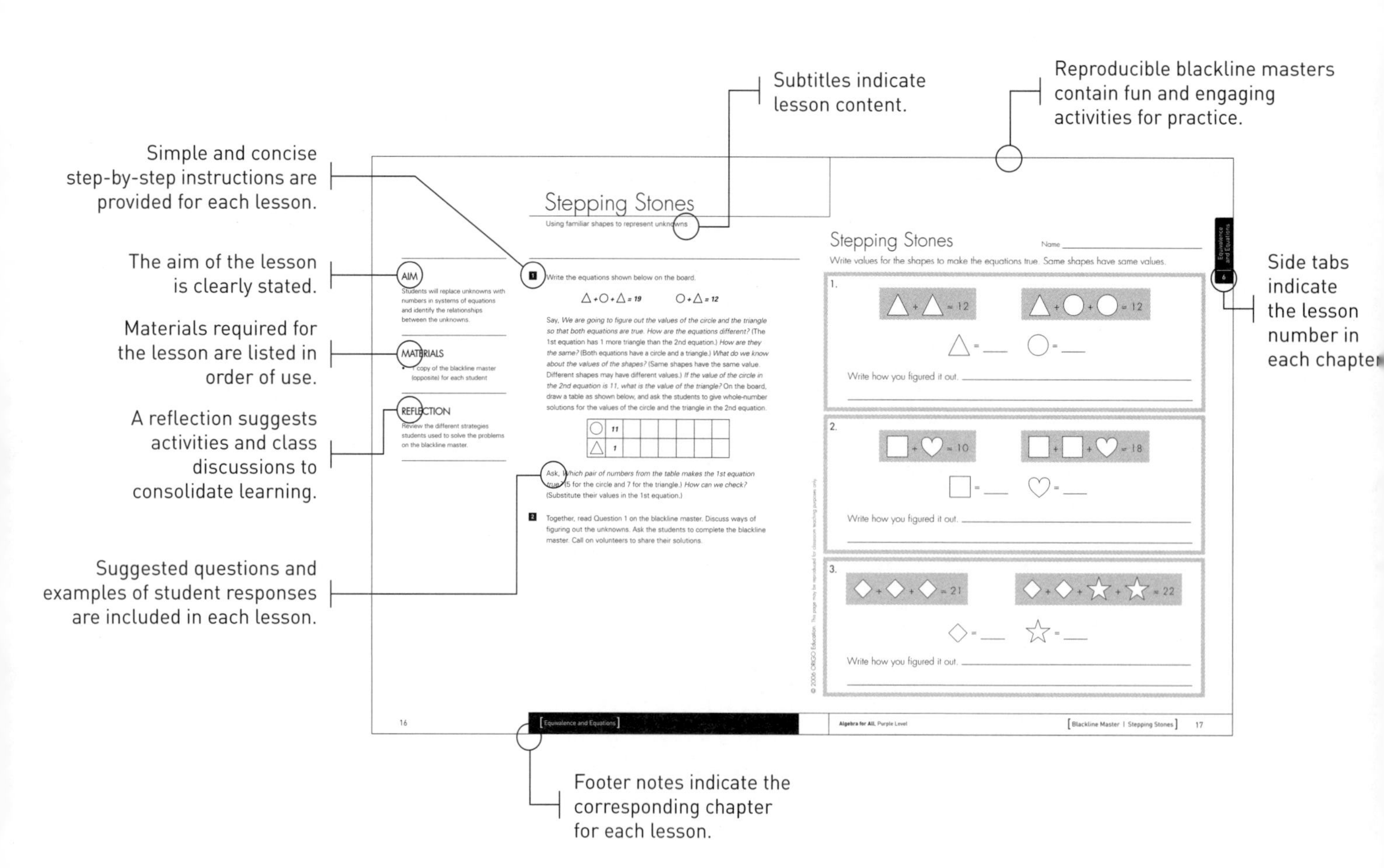

Assessment

Students' thinking is often best gauged by the conversations that occur during classroom discussions. Listen to your students and make notes about their thinking. You may decide to use the rubric below to assess students' mathematical proficiency in the tasks for each lesson. Study the criteria, then assess and record each student's understanding on a copy of the Assessment Summary provided on page 74. Although the summary lists every lesson in this book, it is not necessary to assess students for all lessons.

A	The student fully accomplishes the purpose of the task. Full understanding of the central mathematical ideas is demonstrated. The student is able to communicate his/her thinking and reasoning.
B	The student substantially accomplishes the purpose of the task. An essential understanding of the central mathematical ideas is demonstrated. The student is generally able to communicate his/her thinking and reasoning.
C	The student partially accomplishes the purpose of the task. A partial or limited understanding of the central mathematical ideas is demonstrated and/or the student is unable to communicate his/her thinking and reasoning.
D	The student is not able to accomplish the purpose of the task. Little or no understanding of the central mathematical ideas is demonstrated and/or the student's communication of his/her thinking and reasoning is vague or incomplete.

In the Bag

Using balance to solve addition equations with unknowns

AIM

Students will use the balance method to solve addition problems with one unknown.

MATERIALS

- 1 set of balance scales
- Connecting cubes
- 3 identical non-transparent bags
- 1 copy of the blackline master (opposite) for each student

REFLECTION

Ask, *How can we solve equations that have unknowns?* (Use balancing and inverses.) *How do we check our answers?* (Substitute the solution in the equation.) Call on volunteers to share their thinking. Reinforce the idea that cubes were removed from both sides to reverse addition.

1 Write $6 + 3 = 4 + 5$ on the board. Call on a volunteer to model the equation on the scales with 6 red, 3 green, 4 blue, and 5 yellow cubes. Ask, *Is this an equation? How do you know?* (It has an equals symbol and it balances.) Add 2 cubes to one side of the scales and ask, *Is this equation still balanced? How do we make it balance?* (Add 2 cubes to the other side.) Balance the scales, then remove 3 cubes from one side and repeat the discussion. Discuss how an equation remains balanced if we add or subtract the same amount from each side.

2 Without being observed, place 4 cubes in one of the bags and model "3 add unknown (bag) equals 7" on the scales. Record $3 + \square = 7$ on the board. Ask, *What can we do to both sides of the scales to help us figure out the value of the unknown?* (Take 3 cubes from each side.) Invite a volunteer to remove 3 cubes from each side of the scales. Ask, *Are the scales still balanced?* Write $\square = 4$ on the board and validate the answer by substituting "4" in the equation. Open the bag and show the 4 cubes. Repeat for $\square + 8 = 12$.

3 Show the 3 identical bags. Ask, *How do you know that these bags represent the same unknown?* (They are identical.) Write $\square + \square + 5 = \square + 8$ on the board. Without being observed, place 3 cubes in each bag and model the equation on the scales. Ask, *How can we figure out the value of the unknown?* (Take 5 cubes from each side and then take one unknown from each side.) *Are the scales still balanced? What is the value of the unknown?* (3) Validate the answer by substituting "3" in the equation. Open the bags to show that each holds 3 cubes.

4 Complete the blackline master with the class.

In the Bag

Name ____________________

Solve these equations. Cross out the same amount on each side to keep the scales balanced.

1.

$\_\_\_\_ + 8 = 13$

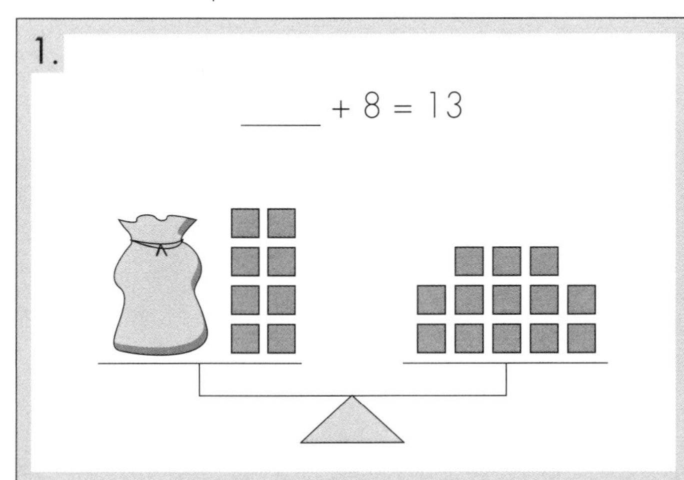

a. How many squares did you cross out on each side? ______

b. Complete the equation.

2.

$14 = \_\_\_\_ + 5$

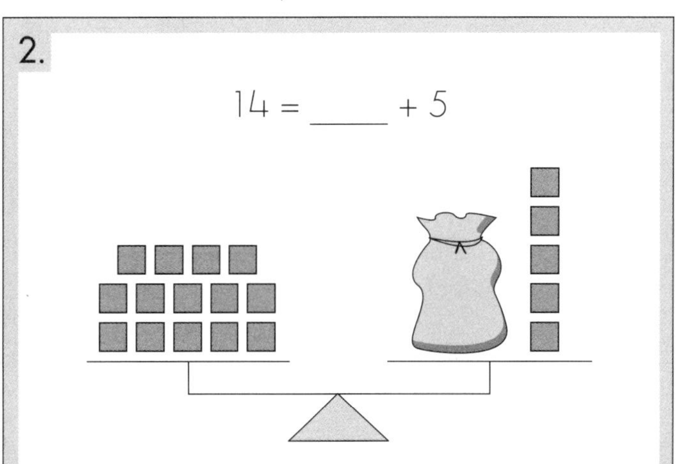

a. How many squares did you cross out on each side? ______

b. Complete the equation.

3.

$\_\_\_\_ + \_\_\_\_ = \_\_\_\_ + 7$

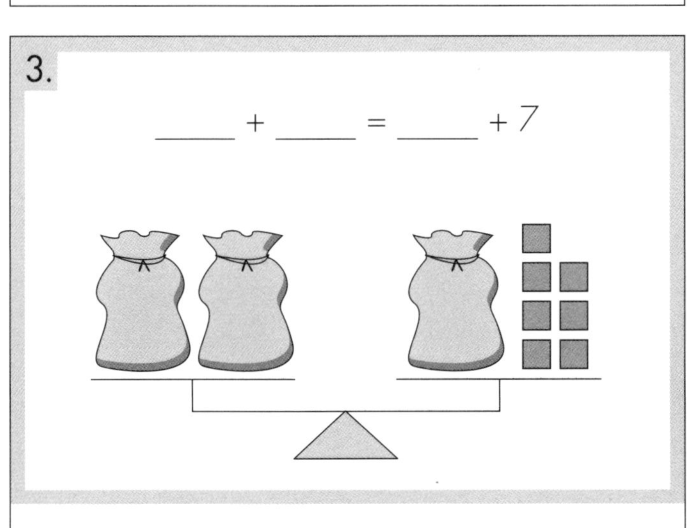

a. What did you cross out on each side?

____

____

b. Complete the equation.

4.

$\_\_\_\_ + \_\_\_\_ + 6 = \_\_\_\_ + 12$

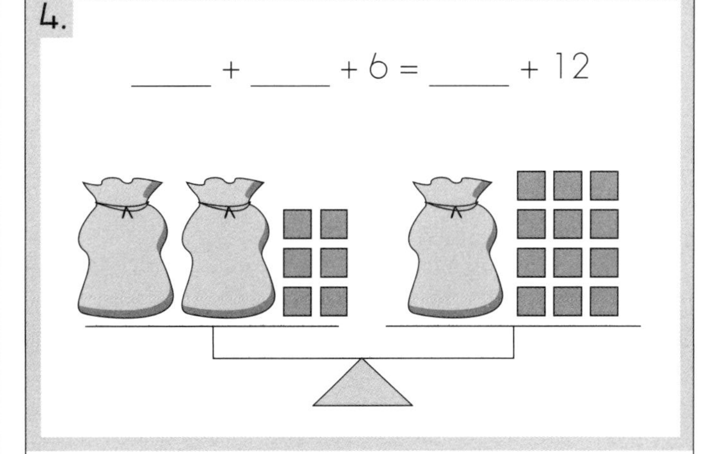

a. What did you cross out on each side?

____

____

b. Complete the equation.

Fair Shares

Using balance to solve equations with the same unknown

AIM

Students will use the balance method to solve equations with one unknown.

MATERIALS

- Connecting cubes for each student
- 1 copy of the blackline master (opposite) for each student

REFLECTION

Ask, *How can we solve equations that have the same unknown?* Encourage students to explain how they could share or use their knowledge of multiplication. Ask, *How do we check our answers?* (Substitute the numbers in the equation.) Call on volunteers to share their thinking.

1 Draw the diagram shown below and write □ + □ **= 16** on the board.

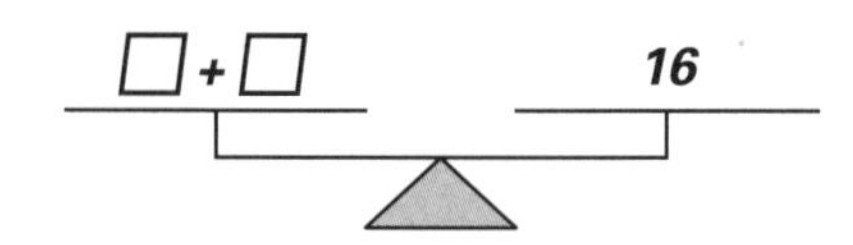

Ask, *What do you know about the missing numbers?* (The same number must be in each □.) *How can we figure out the value of the unknown?* (Figure out the number that doubles to make 16, or share 16 between 2.) Provide cubes for volunteers to show their thinking.

2 Validate the answer by substituting "8" in the equation. Repeat for □ + □ **= 28** and then □ + □ + □ **= 36**. During the discussion, encourage confident individuals to explain that the equations could be written as ***2 ×*** □ ***= 28*** or ***3 ×*** □ ***= 36***.

3 Complete the blackline master with the class.

Fair Shares

Name ____________________

Solve the equations. For each question, draw the same number of squares in each bag to keep the scales balanced. Write the numbers.

1.

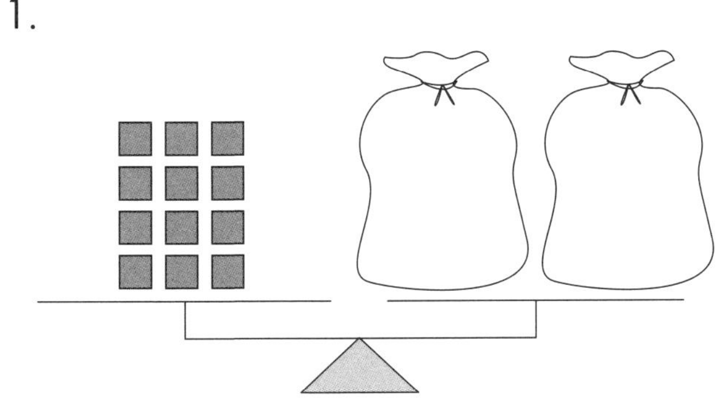

12 = ____ + ____

2.

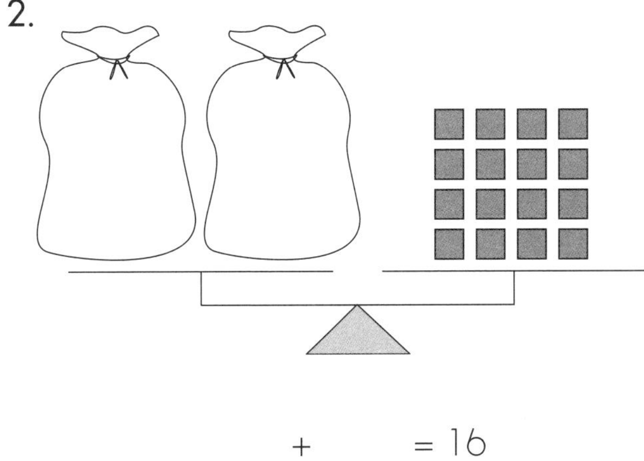

____ + ____ = 16

3.

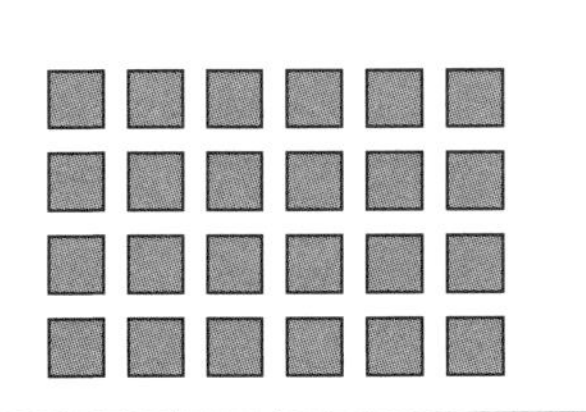

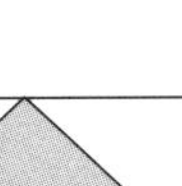

24 = ____ + ____ + ____

4.

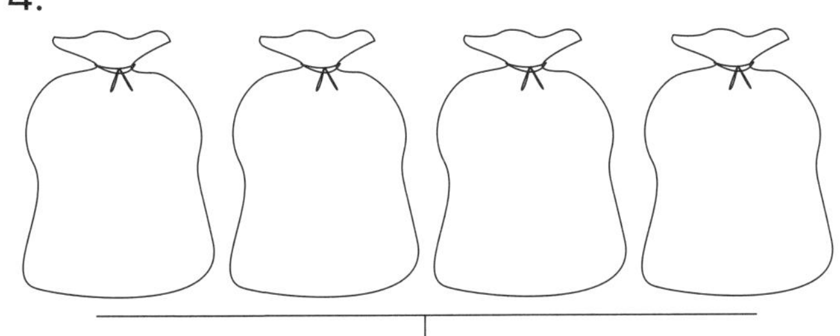

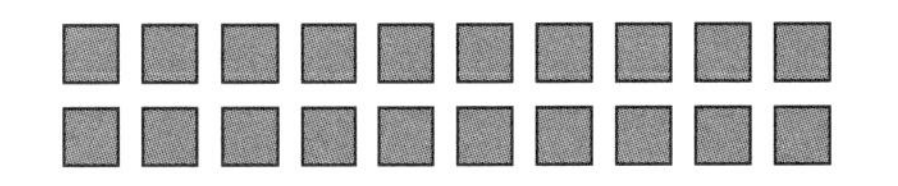

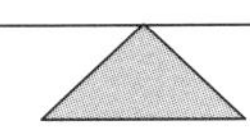

____ + ____ + ____ + ____ = 20

3D Fun

Solving for unknowns in systems of equations

AIM

Students will replace unknowns with numbers in systems of equations and identify the relationships between the unknowns.

MATERIALS

- 1 copy of the blackline master (opposite) for each student

REFLECTION

Review how to solve these problems. (Look at the set of scales with only one type of shape on it and figure out how much it weighs. Find the set of scales with this shape and only one other type of shape, then substitute how much the 1st shape weighs to figure out how much the 2nd shape weighs, and so on.)

1 Refer to Question 1 on the blackline master. Remind the students that same items weigh the same amount, and that different items may weigh the same, or they may weigh different amounts. Ask, *How can we figure out how much each shape weighs? What shapes are on the 1st set of scales? How much do they weigh in total?* (8 kg.) *How much does 1 cone weigh?* (4 kg.) *How can we use this to figure out how much the other 2 shapes weigh? Look at the 3rd set of scales. We know that the cone weighs 4 kg. How much do the rectangular prism and the cone weigh in total?* Write ***Rectangular prism + 4 = 6*** on the board. Ask, *How much does the rectangular prism weigh?* (2 kg.) *Look at the 2nd set of scales. We know that the rectangular prism weighs 2 kg.* Write ***Sphere + 2 = 8*** on the board. Ask, *How much does the sphere weigh?* (6 kg.)

2 Refer to Question 2. Ask questions such as, *What shapes are on the 1st set of scales? What is the total showing on the 2nd set of scales? How much do the cylinder and the triangular prism weigh in total?* Ask the students to complete the blackline master. Call on several volunteers to share how they solved Question 3.

3D Fun

Name ________________________________

Write how much each shape weighs. Same shapes weigh the same.

1.

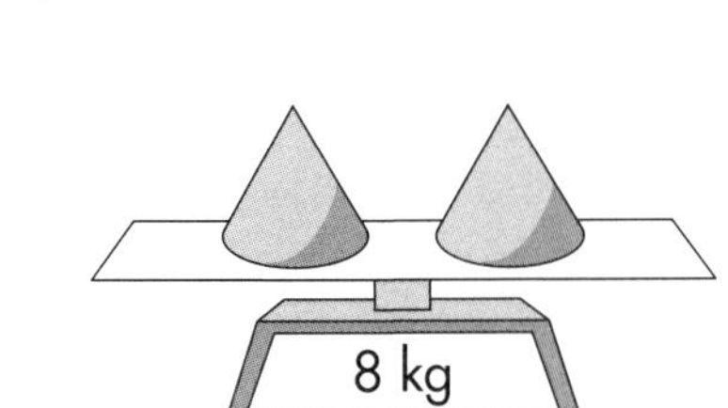

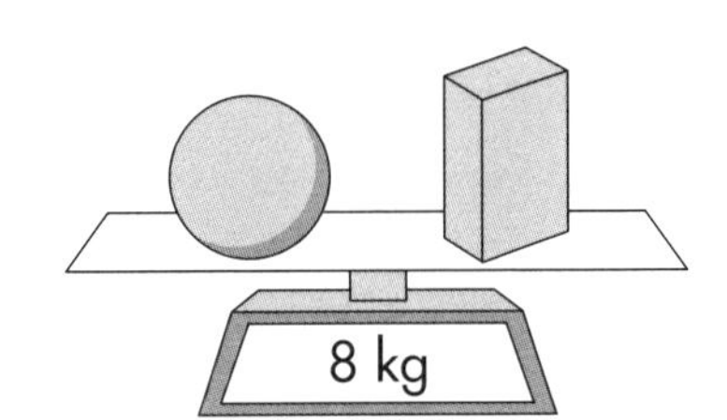

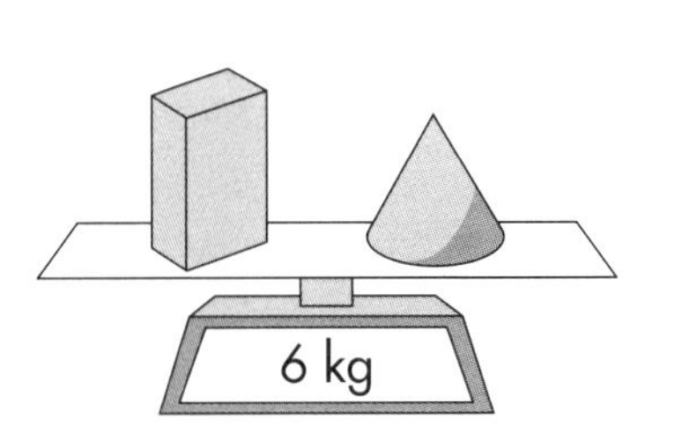

= ________ kg   = ________ kg   = ________ kg

2.

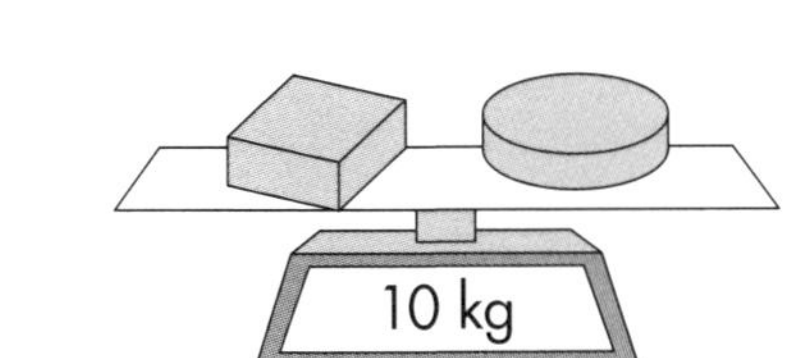

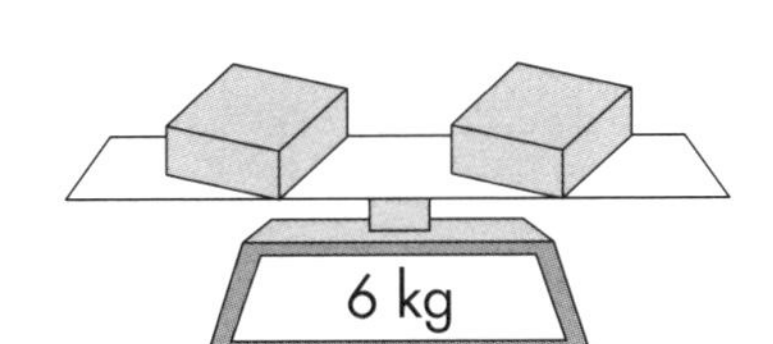

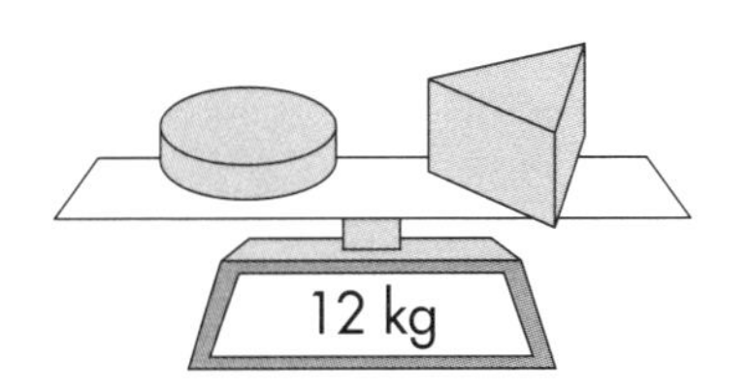

= ________ kg   = ________ kg   = ________ kg

3.

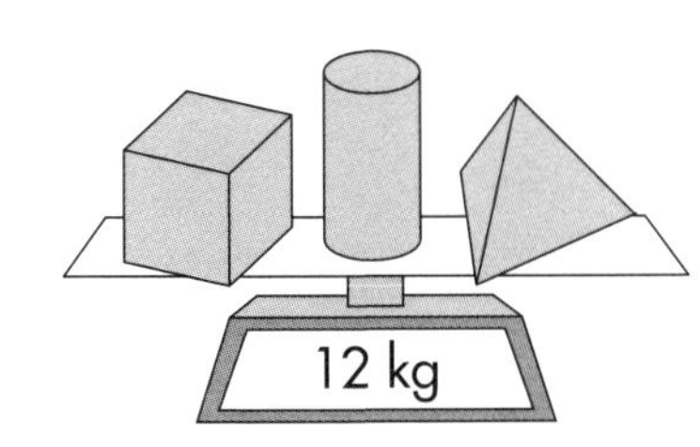

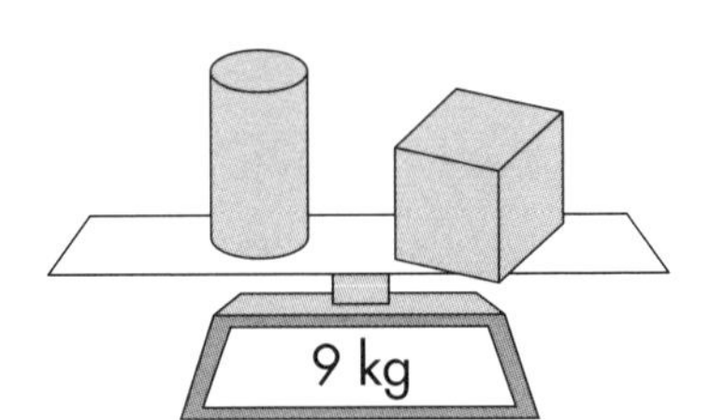

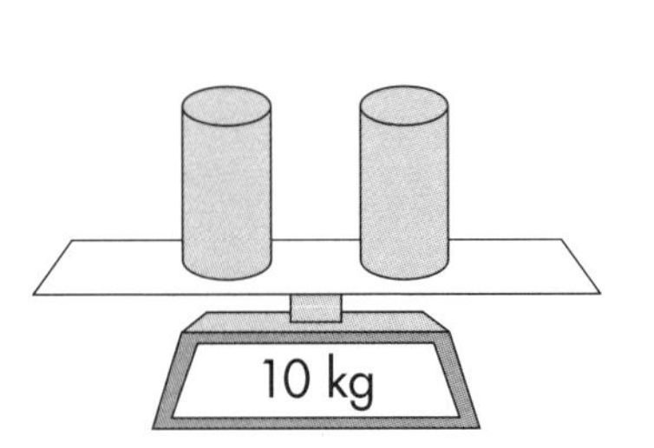

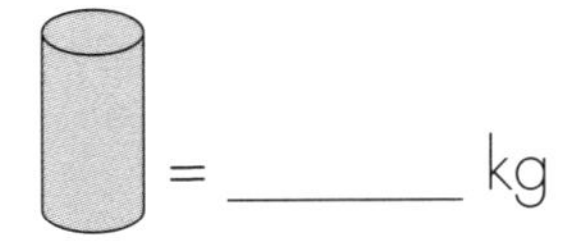 = ________ kg

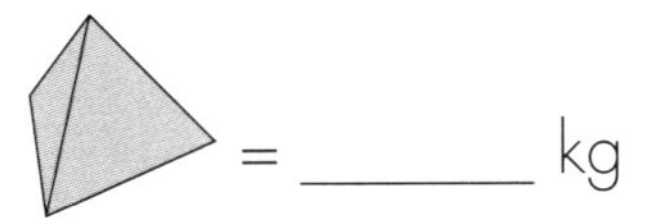 = ________ kg

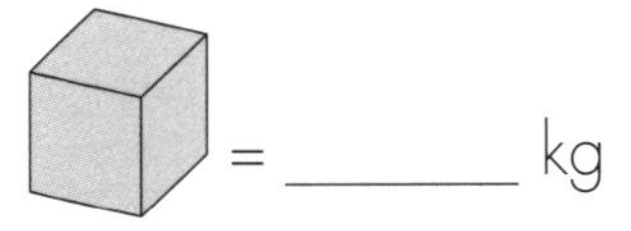 = ________ kg

Liquid Measures

Solving for unknowns in systems of equations

AIM

Students will replace variables with numbers in systems of equations and identify the relationships between the variables.

MATERIALS

- 1 copy of the blackline master (opposite) for each student

REFLECTION

Review the different ways of solving the problems on the blackline master.

1 On the board, draw the diagrams shown below.

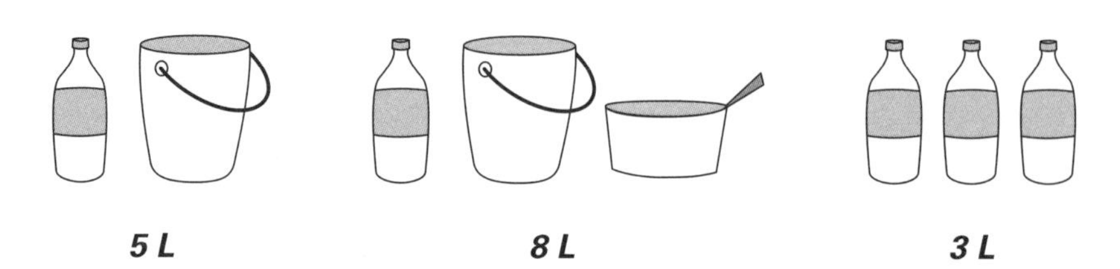

Ask, *How can we figure out how much water each container holds?* Call on volunteers to suggest ways of solving the problem. Discuss the following solution.

- If 3 bottles together hold 3 L, then each bottle must hold 1 L.
- If one bottle and the bucket together hold 5 L, then the bucket must hold 4 L.
- If the bottle, the bucket, and the saucepan together hold 8 L, then the saucepan must hold 3 L.

2 Refer to Question 1 on the blackline master. Ask questions such as, *How much water do the 3 saucepans hold in total? How can we figure out how much the bucket holds? How can we figure out how much the bottle holds?* Ask the students to complete the blackline master. Call on volunteers to share their answers to Questions 2 and 3.

Liquid Measures

Name ______________________________

Write how much each container holds. Same containers hold the same.

1.

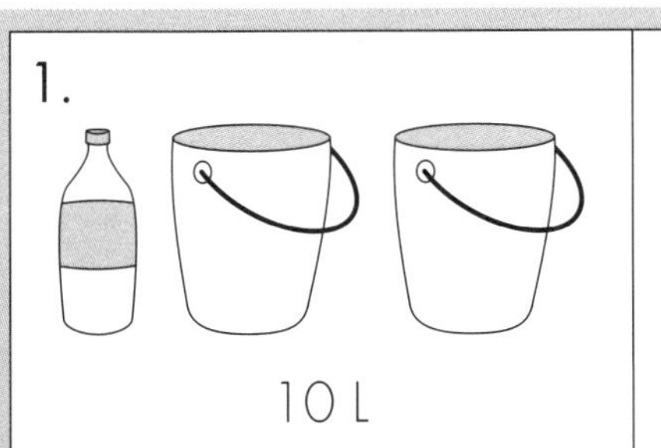

10 L

7 L

9 L

 = ________ L

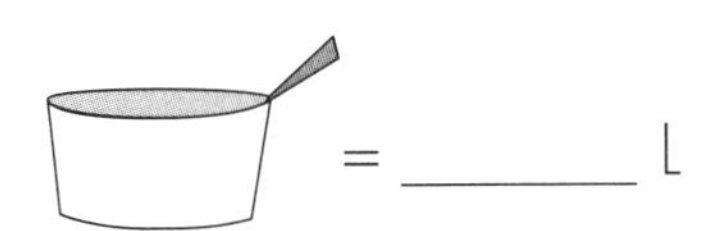 = ________ L

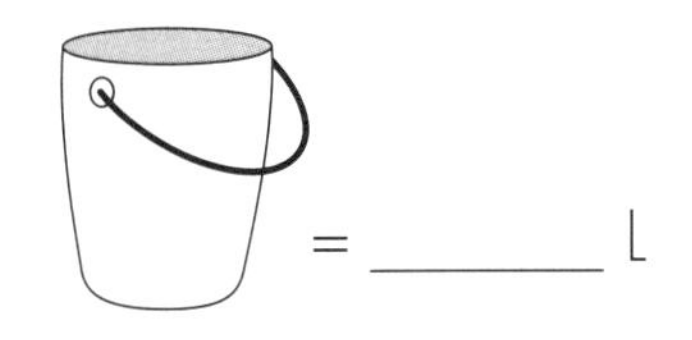 = ________ L

2.

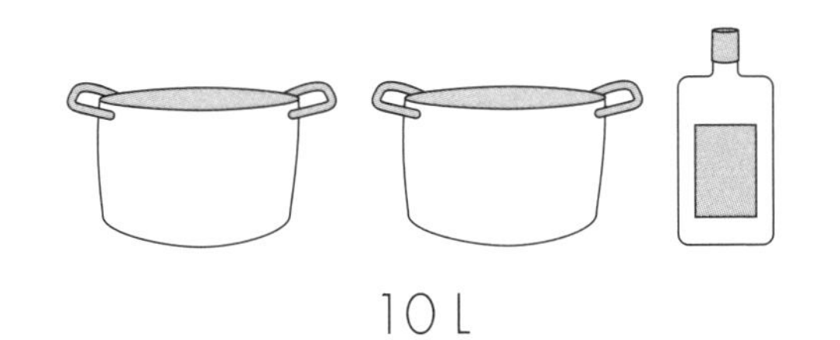

10 L

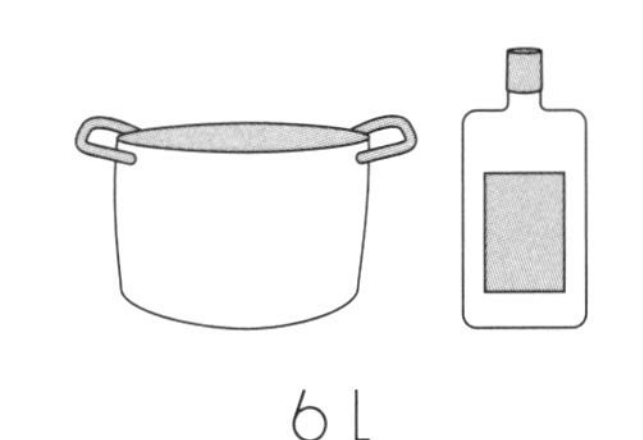

6 L

 = ________ L

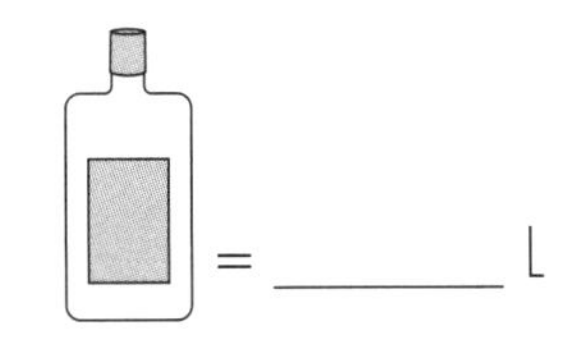 = ________ L

3.

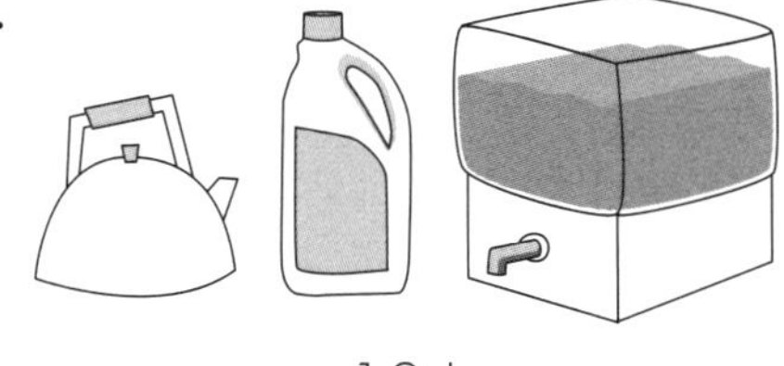

10 L

5 L

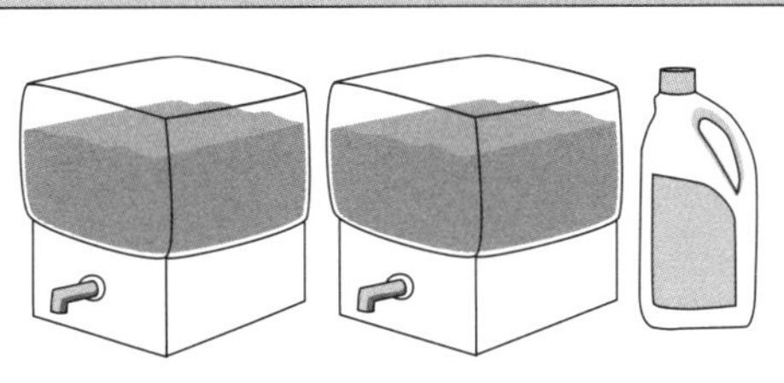

12 L

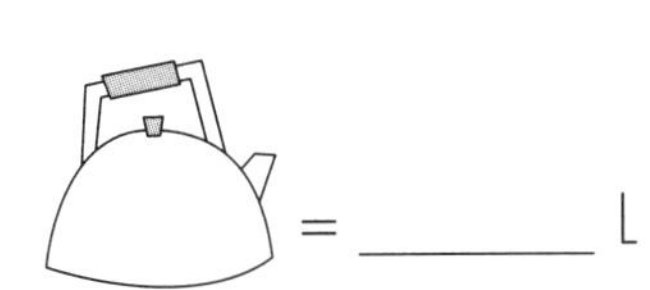 = ________ L

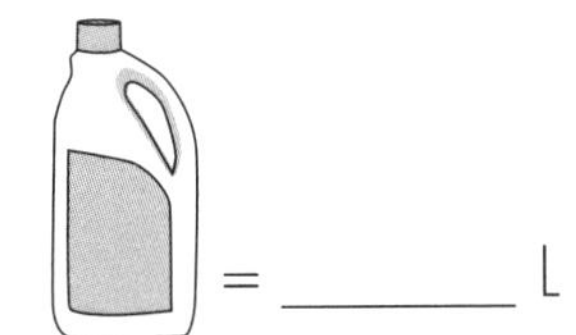 = ________ L

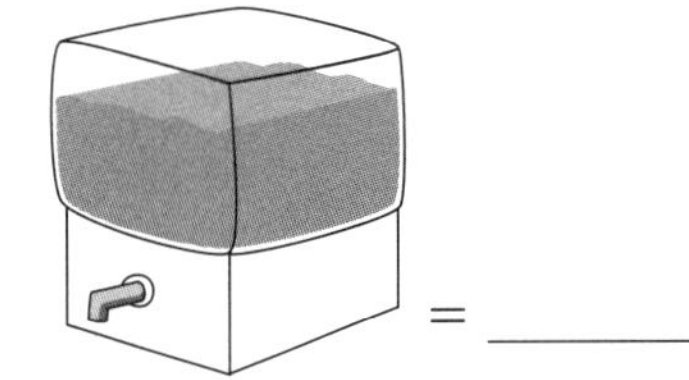 = ________ L

Number Stories

Using scales to represent "less than" and "greater than"

AIM

Students will match numbers to "greater than" or "less than" situations.

MATERIALS

- 1 copy of the blackline master (opposite) for each student

REFLECTION

Discuss 3 ways to describe the relationship between 2 numbers that are not equal. For example, for 23 and 34:

- 23 is not equal to 34
- 23 is less than 34
- 34 is greater than 23.

1 Ask, *Will has 3 trucks and 2 cars, and Josh has 5 trucks and 1 car. Do they have the same number of toys? How do we write this?* Discuss the symbols "=" and "≠". Write ***3 + 2 ≠ 5 + 1*** on the board. Call on volunteers to tell other "unequal" stories. Each time, write the matching number sentence on the board.

2 Say, *3 + 2 is not equal to 5 + 1. Which sum is "greater than" and which sum is "less than"? Imagine each of the cars weigh the same and each of the trucks weigh the same. What will 3 + 2 ≠ 5 + 1 look like on a set of balance scales? Which side will be heavier? Which side will be lighter?* On the board, draw the diagrams shown below.

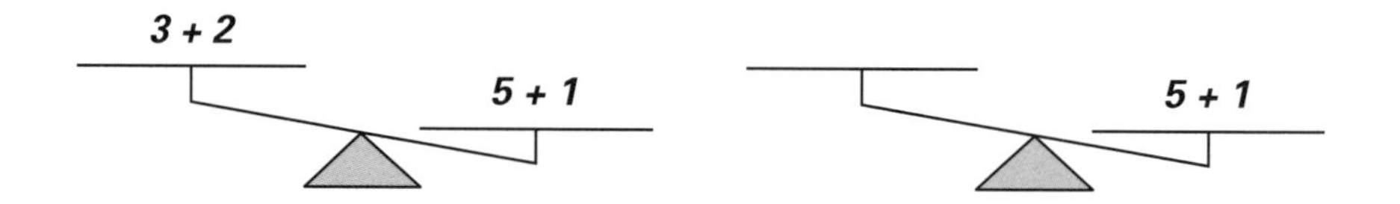

3 Indicate the left side of the 2nd set of scales and ask, *What number can we write here to make this true*? Allow time for volunteers to suggest different numbers. Discuss how any number less than 6 will make the balance scales true.

4 Ask the students to complete the blackline master. Call on volunteers to share their responses.

Number Stories

Name ______________________________

1. Write a number to make each true. Use = or ≠ to write a matching number sentence.

a.

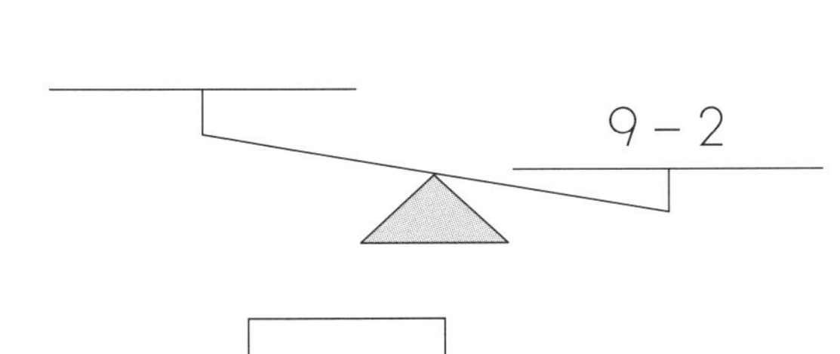

____ ≠ 9 − 2

b.

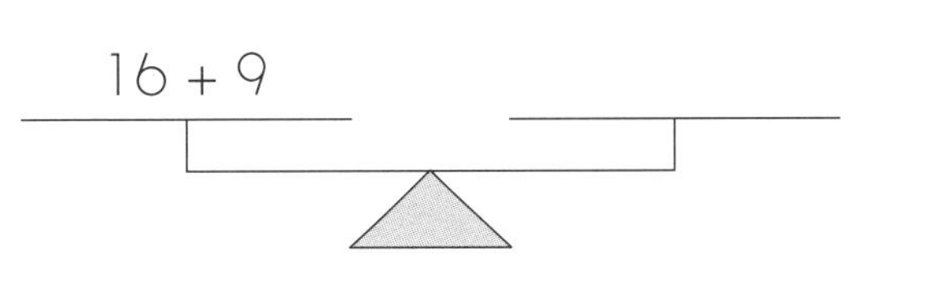

____

c.

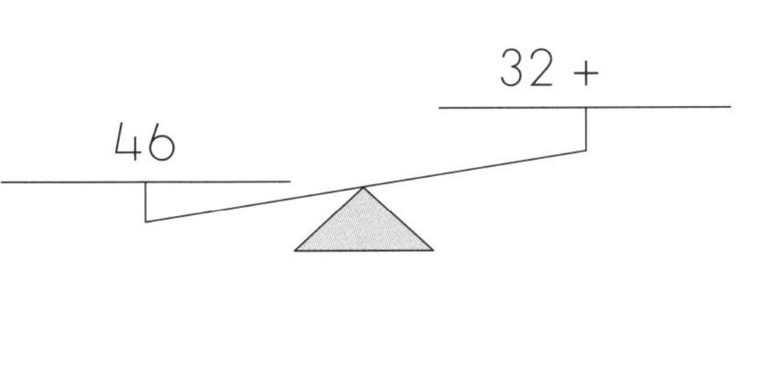

____

d.

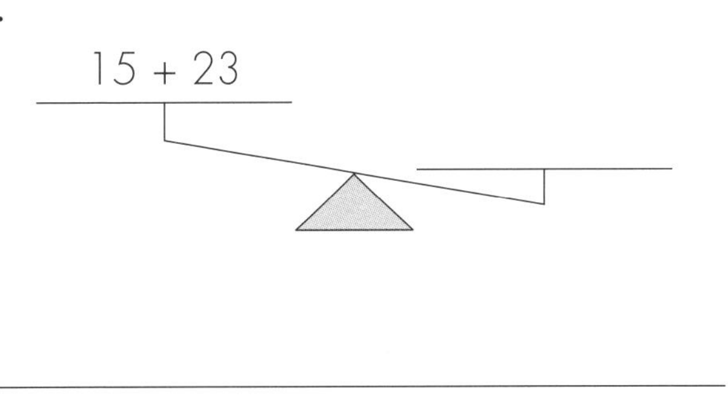

____

2. Write numbers to make these true. Use = or ≠ to write a matching number sentence.

a.

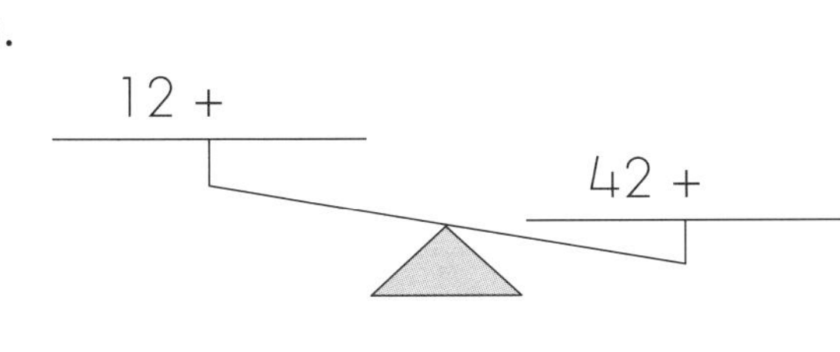

12 + ____ ≠ 42 + ____

b.

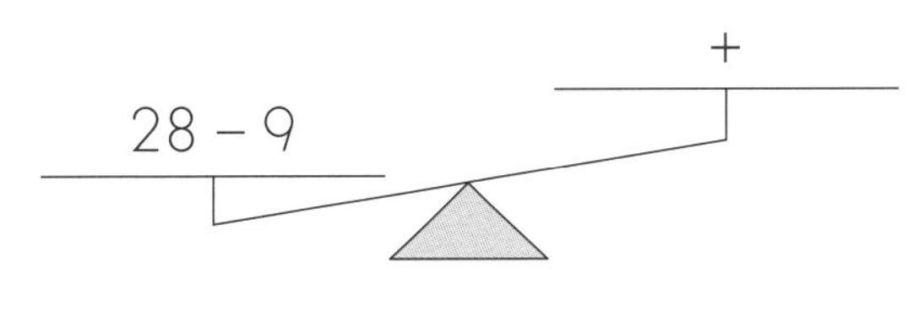

____

c.

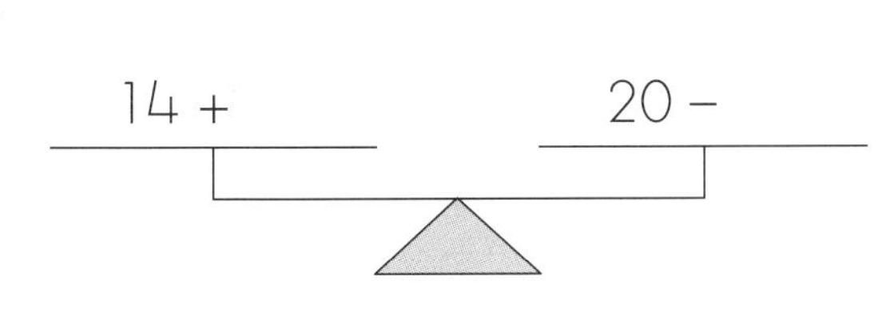

____

d.

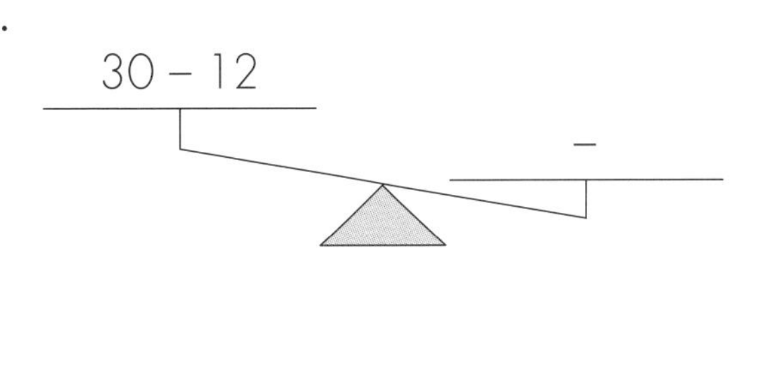

____

Stepping Stones

Using familiar shapes to represent unknowns

AIM

Students will replace unknowns with numbers in systems of equations and identify the relationships between the unknowns.

MATERIALS

- 1 copy of the blackline master (opposite) for each student

REFLECTION

Review the different strategies students used to solve the problems on the blackline master.

1 Write the equations shown below on the board.

$\triangle + \bigcirc + \triangle = 19 \qquad \bigcirc + \triangle = 12$

Say, *We are going to figure out the values of the circle and the triangle so that both equations are true. How are the equations different?* (The 1st equation has 1 more triangle than the 2nd equation.) *How are they the same?* (Both equations have a circle and a triangle.) *What do we know about the values of the shapes?* (Same shapes have the same value. Different shapes may have different values.) *If the value of the circle in the 2nd equation is 11, what is the value of the triangle?* On the board, draw a table as shown below, and ask the students to give whole-number solutions for the values of the circle and the triangle in the 2nd equation.

○	**11**						
△	**1**						

Ask, *Which pair of numbers from the table makes the 1st equation true?* (5 for the circle and 7 for the triangle.) *How can we check?* (Substitute their values in the 1st equation.)

2 Together, read Question 1 on the blackline master. Discuss ways of figuring out the unknowns. Ask the students to complete the blackline master. Call on volunteers to share their solutions.

Stepping Stones

Name ______________________________

Write values for the shapes to make the equations true. Same shapes have same values.

1.

△ + △ = 12   △ + ○ + ○ = 12

△ = ____   ○ = ____

Write how you figured it out. ____

____

2.

□ + ♡ = 10   □ + □ + ♡ = 18

□ = ____   ♡ = ____

Write how you figured it out. ____

____

3.

◇ + ◇ + ◇ = 21   ◇ + ◇ + ☆ + ☆ = 22

◇ = ____   ☆ = ____

Write how you figured it out. ____

____

Taller Shorter

Investigating the transitive property

AIM

Students will work with "if then" statements to reason that if ***a*** is greater than ***b***, and ***b*** is greater than ***c***, then ***a*** is greater than ***c***.

MATERIALS

- 1 copy of the blackline master (opposite) for each student

REFLECTIONS

On the board, write "if then" statements for each of the examples on the blackline master. Invite volunteers to complete the statements using the information shown along each number line. For example, ***If (Gemma) is taller than (Chi), and (Chi) is taller than (Luisa), then (Gemma) is taller than (Luisa).***

1 Ask, *If Len is 134 cm tall, and Dee is 150 cm tall, is Len taller than Dee?* (No. Len is shorter than Dee.) *How much shorter?* (150 – 134 = 16) *Kara is 129 cm tall. Is she taller than Dee?* (No. Dee is taller than Kara.) *How much taller is Dee?* (150 – 129 = 21) *How much taller than Kara is Len?* (134 – 129 = 5) Draw an open number line on the board. Invite volunteers to label points along the line to show the relative position of each person, as shown below.

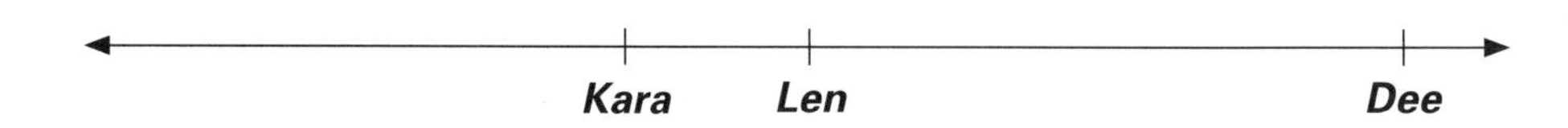

Record this "if then" statement on the board: ***If Len is taller than Kara, and Dee is taller than Len, then Dee is taller than Kara.*** Ask the students to complete Question 1 on the blackline master. Call on volunteers to share their answers.

2 Say, *Oliver is 5 cm taller than Brad, and Brad is 3 cm taller than Leona. How much taller than Leona is Oliver? How can we figure it out?* (Draw an open number line on the board and model the problem.) Mark Leona's height and ask, *Where do we mark Brad's height?* (3 cm greater than Leona.) *Where do we mark Oliver's height?* (5 cm greater than Brad.) *What is the difference between Leona's height and Oliver's height?* (8 cm.)

Record this "if then" statement on the board: ***If Oliver is 5 cm taller than Brad, and Brad is 3 cm taller than Leona, then Oliver is 8 cm taller than Leona.***

3 Ask the students to complete the blackline master.

Taller Shorter

Name ______________________________

For each of these, write names on the number line to show the position of each person. Then complete the sentences.

1. Gemma is 5 cm taller than Chi. Chi is 7 cm taller than Luisa.

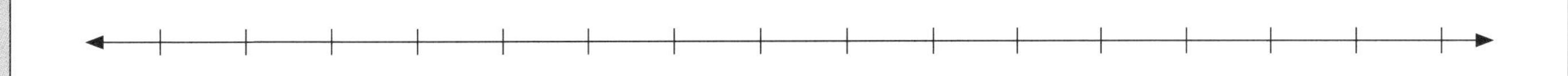

   a. ______________ is the shortest.

   b. Gemma is _______ cm taller than Luisa.

2. Frank is 8 cm taller than Alex. Alex is 3 cm shorter than Carlos.

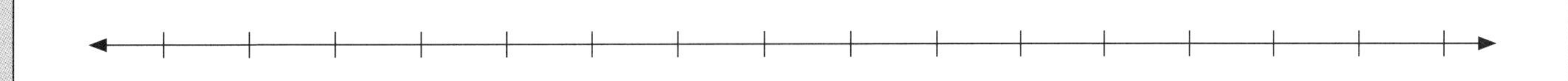

   a. ______________ is the shortest.

   b. Frank is _______ cm taller than Carlos.

3. Jacinta is 11 cm shorter than Grace. Grace is 5 cm taller than Meg.

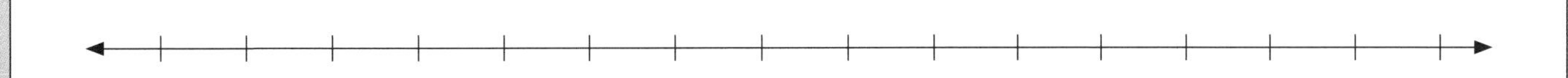

   a. ______________ is the tallest.

   b. Meg is _______ cm taller than Jacinta.

Older Younger

Investigating the transitive property

AIM

Students will work with "if then" statements to reason that if ***a*** is greater than ***b***, and ***b*** is greater than ***c***, then ***a*** is greater than ***c***.

MATERIALS

- 1 copy of the blackline master (opposite) for each student

REFLECTION

On the board, write "if then" statements for each of the examples on the blackline master. Invite volunteers to complete the statements using the information shown along each number line. For example, ***If (Anita) is older than (Leah), and (Leah) is older than (Jon), then (Anita) is older than (Jon).***

1 Say, *Kym is 5 years older than Lee. Kym is 9 years older than Elise. Who is older, Lee or Elise? How do you know?* On the board, draw an open number line and invite students to label points along the line to show the relative positions of the 3 people, as shown below.

Write this "if then" statement on the board: ***If Elise is younger than Lee, and Lee is younger than Kym, then ______ is younger than ______.*** Ask, *How can we write an "older than" statement about Kym, Elise, and Lee?* On the board, write, ***If Kym is older than Lee, and Lee is older than Elise, then ______ is older than ______.*** Invite volunteers to complete the statements.

2 Ask the students to complete the blackline master.

Older Younger

Name ______________________

For each of these, write names on the number line to show the position of each person. Then complete the sentences.

1. Anita is 5 years older than Leah. Leah is 6 years older than Jon.

a. ______________ is the youngest.

b. Anita is _______ years older than Jon.

2. Justin is 12 years younger than Tarja. Tarja is 4 years older than Kerin.

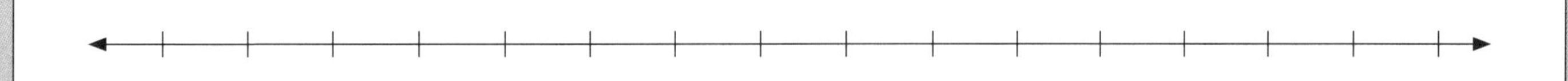

a. ______________ is the oldest.

b. Kerin is _______ years older than Justin.

3. Gina is 2 years younger than Tomas. Tomas is 14 years older than Tara.

a. ______________ is the youngest.

b. Tara is _______ years younger than Gina.

Growing Triangles

Predicting parts in growing patterns

AIM

Students will identify growing patterns and record the data in a table. They will also predict parts in the pattern.

MATERIALS

- Pattern blocks
- 7 signs: "Picture 1", "Picture 2", "Picture 3", "Picture 4", "Picture 5", "Picture 6", and "Picture 10"
- 1 copy of the blackline master (opposite) for each student

REFLECTION

Refer to the blackline master and ask, *How did you figure out the pattern rules?* Call on volunteers to share their ideas.

1 Seat the students on the floor. Use the blocks to make the following pattern. Place the matching sign below each part, as shown.

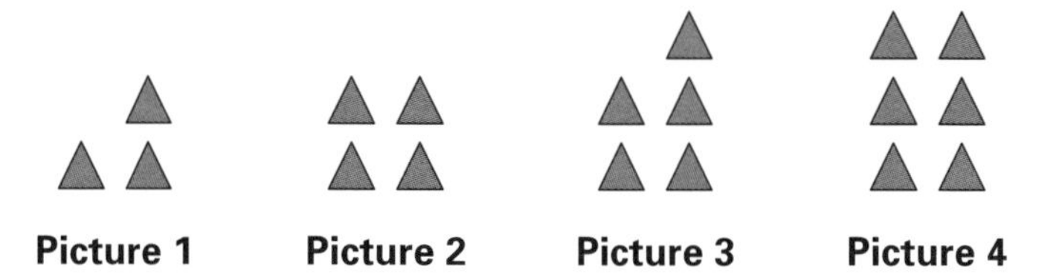

Ask, *How is the pattern growing? What is happening to the 1st column in each picture? What is happening to the 2nd column? How can we use the picture number to help us figure out the number of blocks in each picture? What will Picture 5 look like?* Call on volunteers to make Pictures 5, 6, and 10.

2 On the board, draw the table shown below. Ask, *How many blocks are in Picture 1?* Repeat for each picture. Record the answers in the table.

Picture	Number of blocks

Ask, *What patterns can you see in the table?* (Both columns in the table add 1 each picture. For each picture, there are 2 more blocks than the picture number.) *How many blocks will be in Picture 20?* (22)

3 Ask the students to complete the blackline master. Call on volunteers to share their answers.

Growing Triangles

Name ________________________

1. a. Draw the next 2 pictures in this pattern.

 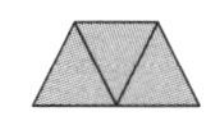 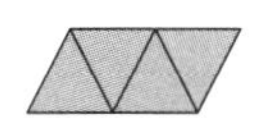 

Picture 1 Picture 2 Picture 3 Picture 4 Picture 5 Picture 6

b. Complete this table.

Picture	1	2	3	4	5	6	10	20
Number of triangles								

c. How did you figure out the number of triangles in Picture 10?

____

____

2. a. Draw the next 2 pictures.

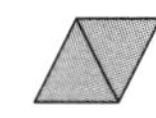 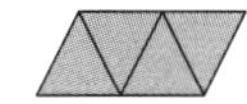 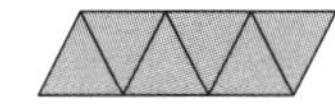 

Picture 1 Picture 2 Picture 3 Picture 4 Picture 5 Picture 6

b. Complete this table.

Picture	1	2	3	4	5	6	10	20
Number of triangles								

c. How did you figure out the number of triangles in Picture 10?

____

____

Growing Squares

Predicting part numbers in growing patterns and writing pattern rules

AIM

Students will identify growing patterns and record the data in a table. They will also predict parts in the pattern and write the pattern rule.

MATERIALS

- Identical tiles or pattern blocks
- 7 signs: "Picture 1", "Picture 2", "Picture 3", "Picture 4", "Picture 5", "Picture 6", and "Picture 10"
- 1 copy of the blackline master (opposite) for each student

REFLECTION

Refer to the blackline master. Discuss the pattern rule for each question. Ask, *How did you figure out the pattern rule?* Invite several volunteers to share their thinking.

1 Seat the students on the floor. Use the tiles to make the following pattern. Place the appropriate sign below each picture, as shown.

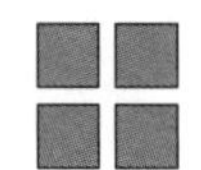

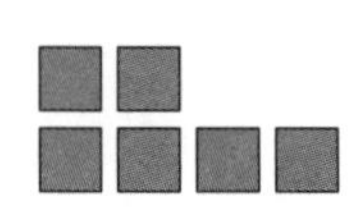
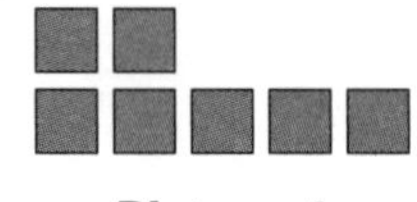

Picture 1 **Picture 2** **Picture 3** **Picture 4**

Ask, *How is the pattern growing? What is happening to the 1st row? What is happening to the 2nd row? How can we use the picture number to help us figure out the number of tiles in each picture? What will Picture 5 look like?* Invite volunteers to make Pictures 5, 6, and 10.

2 On the board, draw the table shown below. Ask, *How many tiles are in Picture 1?* Repeat for each picture. Record the answers in the table.

Picture	Number of tiles

Ask, *What pattern can you see in the table?* (The number of tiles is 3 more than the picture number.) *How many tiles will be in Picture 20? If there are 33 tiles in a picture, what would be the picture number?* (30) *What is the pattern rule?* (The number of tiles = the picture number add 3.)

3 Ask the students to complete the blackline master. Call on volunteers to share their answers.

Growing Squares

Name ______________________________

1. a. Draw the next 2 pictures in this pattern.

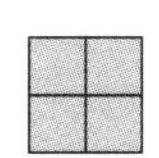 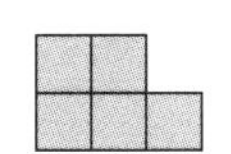 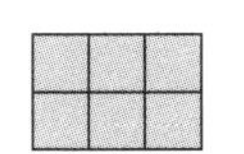 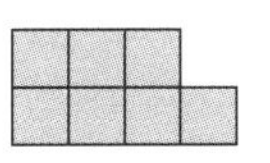

Picture 1 | Picture 2 | Picture 3 | Picture 4 | Picture 5 | Picture 6

b. Complete this table.

Picture	1	2	3	4	5	6	10	20
Number of squares								

c. How did you figure out the number of squares in Picture 10?

____

____

2. a. Draw the next 2 pictures.

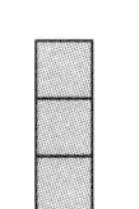 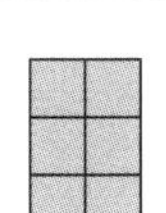 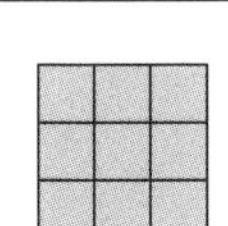 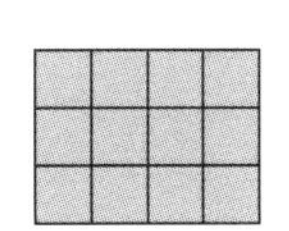

Picture 1 | Picture 2 | Picture 3 | Picture 4 | Picture 5 | Picture 6

b. Complete this table.

Picture	1	2	3	4	5	6	10	20
Number of squares								

c. How did you figure out the number of squares in Picture 10?

____

____

Pick the Part

Identifying the relationship between numbers in a growing pattern and the position of the numbers

AIM

Students will identify growing patterns and the relationship between numbers in a growing pattern, then use the relationship to extend the pattern.

MATERIALS

- Red and blue tiles
- 7 labels: "Picture 1", "Picture 2", "Picture 3", "Picture 4", "Picture 5", "Picture 6", and "Picture 10"
- 1 copy of the blackline master (opposite) for each student

REFLECTION

Refer to the blackline master and discuss the growing pattern in Question 2. Ask, *How did you figure out the number of triangles? How did you figure out the number of squares?* Call on volunteers to share their reasoning.

1 Seat the students on the floor. Use the tiles and labels to make the following pattern.

2 Ask, *How many red and blue tiles are in Picture 1?* (2 red and 2 blue tiles.) *How many red and blue tiles are in Picture 2?* (3 red and 4 blue tiles.) *How many are in Picture 3?* (4 red and 6 blue tiles.) *How many red and blue tiles will be in Picture 4? How do you know?* Discuss how the number of red tiles is always 1 more than the picture number, and the number of blue tiles is always double the picture number. Call on a volunteer to arrange the red tiles for Picture 4. Repeat the discussion for Pictures 5 and 6. Ask, *How can we figure out the number of red and blue tiles for Picture 10?* Call on volunteers to explain their thinking.

3 Ask the students to complete the blackline master.

Pick the Part

Name ____________________

1. a. Draw the next 2 pictures in this pattern.

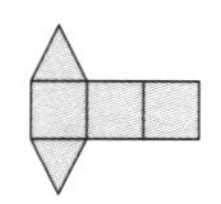

Picture 1

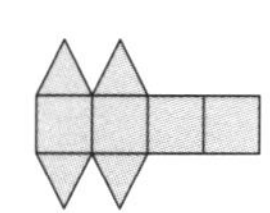

Picture 2

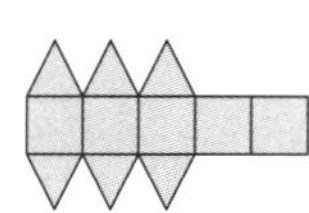

Picture 3

Picture 4

Picture 5

b. Complete this table.

Picture	1	2	3	4	5	10	20
Number of triangles							
Number of squares							

c. How did you figure out the number of triangles in Picture 10?

____

d. How did you figure out the number of squares in Picture 10?

____

2. a. Draw the next 2 pictures in this pattern.

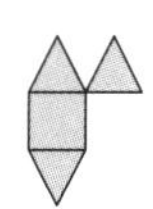

Picture 1

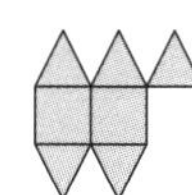

Picture 2

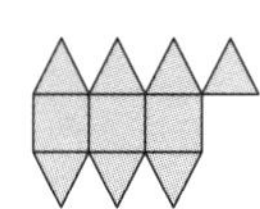

Picture 3

Picture 4

Picture 5

b. Complete this table.

Picture	1	2	3	4	5	10	20
Number of triangles							
Number of squares							

Patterns and Functions 3

Pattern Parts

Identifying and predicting repeating patterns

AIM

Students will examine repeating patterns, record the parts, and predict patterns for a large number of repeats.

MATERIALS

- Red and blue tiles
- 1 copy of the blackline master (opposite) for each student

REFLECTION

Ask, *If we have 100 repeats of the pattern on the blackline master, what would be the total number of shapes? The number of squares? The number of circles? How do you know?*

TEACHING NOTE

Students may try to identify patterns by looking down the columns of the table instead of across the rows. For example, they may say "the total number of tiles is increasing by 3". Try classifying rules as either "down" rules or "across" rules, and remind the students that in this exercise they are searching for "across" rules.

1 Seat the students on the floor. Use the tiles to make 5 parts of the repeating pattern, "red, red, blue, red, red, blue". Ask, *What type of pattern is this?* (A repeating pattern.) Call on a volunteer to separate the pattern into its repeating parts. Ask, *How many tiles are in the 1st repeat?* (3) *How many red tiles are in the 1st repeat?* (2) *How many blue tiles are in the 1st repeat?* (1)

2 On the board, draw the following table.

Repeat number	Total number of tiles	Number of red tiles	Number of blue tiles

Call on a volunteer to point to the first 2 repeats. Move these repeats back together and ask, *How many tiles are in the first 2 repeats?* (6) *How many red tiles are there?* (4) *How many blue tiles are there?* (2) Record the answers in the table. Continue until the table is completed.

3 Ask, *What patterns can you see across the table? What happens to the total number of tiles at each repeat?* (The total is the repeat number multiplied by 3.) *How can we use the repeat number to help figure out the total? How can we figure out the number of red tiles at each repeat?* (It is the repeat number multiplied by 2.) *How can we figure out the number of blue tiles?* (It is the same as the repeat number.) *What will be the total number of tiles in 10 repeats? The number of red tiles? The number of blue tiles?* Repeat for 20 and 100 repeats. See teaching note.

4 Ask the students to complete the blackline master. Call on volunteers to share and explain their responses.

Pattern Parts

Name ______________________

Look at this repeating pattern.

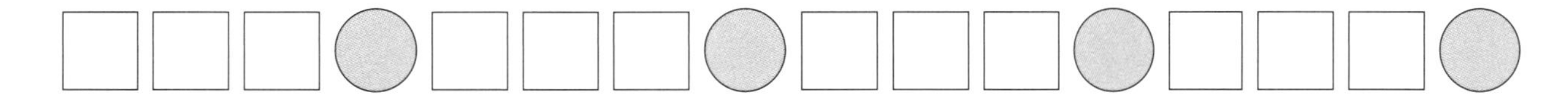

1. Complete this table.

Repeat number	Total number of shapes	Number of squares	Number of circles
1	4	3	1
2			
3			
4			

2. Write how you could use the repeat number to figure out

   a. the total number of shapes.

   ____

   b. the number of squares.

   ____

   c. the number of circles.

   ____

3. Use your answers to Question 2 to complete this table.

Repeat number	Total number of shapes	Number of squares	Number of circles
10			
12			
15			

Patterns and Functions | 4

Making Patterns

Making repeating patterns from pattern parts

AIM

Students will examine parts of repeating patterns and record the number of shapes in each part. They will also make repeating patterns from the 1st repeat.

MATERIALS

- Red and blue tiles
- 1 copy of the blackline master (opposite) for each student

REFLECTION

Discuss how the students figured out the total number of each shape in each pattern on the blackline master.

1 Seat the students on the floor. Draw the following table on the board.

Repeat number	Total number of tiles	Number of red tiles	Number of blue tiles

Arrange tiles into the repeating pattern shown below.

Ask, *What type of pattern is this?* (A repeating pattern.) Call on a volunteer to point to the 1st repeat. Ask, *How many tiles are in the 1st repeat?* (3) *How many red tiles are in the 1st repeat?* (2) *How many blue tiles are in the 1st repeat*? (1) Record the answers in the table. Call on a volunteer to point to the first 2 repeats. Ask, *How many tiles are in the first 2 repeats?* (6) *How many red tiles are in the first 2 repeats?* (4) *How many blue tiles are in the first 2 repeats?* (2) Continue for up to 4 repeats. Then continue the discussion for a pattern with 2 red tiles and 3 blue tiles in each repeat.

2 Ask, *If there are 3 blue tiles and 1 red tile in the 1st repeat, what will the 1st repeat look like? What will 2 repeats look like? What will 5 repeats look like?* Invite volunteers to make the repeating pattern. Ask, *What will be the total number of tiles in 10 repeats? How many red tiles will there be? How many blue tiles? How do you know?* Repeat the discussion for 4 red tiles and 1 blue tile in the 1st repeat.

3 Ask the students to complete the blackline master. In Question 2b, make sure the students understand that the table shows no data for the first 3 repeats.

Making Patterns

Name ______________________

1. Look at this repeating pattern.

□□□□○○○□□□□○○○□□□□○○○

a. Write the missing numbers in this table.

Repeat number	Total number of tiles	Number of squares	Number of circles
1	7	4	3
2			
3			
4			
5			
	70		

2. a. Draw a repeating pattern with 3 squares and 2 circles in the first repeat. Draw 4 repeats.

b. Complete this table for your pattern.

Repeat number	Total number of shapes	Number of squares	Number of circles
10			
12			
15			

Patterns and Functions | 5

Equal Pieces

Using fractions to investigate patterns

AIM

Students will explore patterns when the number of equal parts in one whole and a fraction of that one whole change.

MATERIALS

- 7 sheets of paper
- 1 copy of the blackline master (opposite) for each student

REFLECTION

Invite students to describe rules they can use to figure out the number of shaded parts and the total number of parts from the picture number.

TEACHING NOTE

This activity does not formally use the language of fractions. However, many students will naturally describe the shaded regions using fractions. If the students are confident, a 4th column could be added to the table to record the fraction that is shown in each picture. Some students will observe that the different fractions can be used to describe the same shaded region of the paper.

1 Fold a sheet of paper into 4 equal parts and shade one of the parts as shown in Picture 1 below. Ask questions such as, *How many equal parts can you see? How many of those parts are shaded?* Write "Picture 1" on the sheet of paper. On the board, draw the table shown below and record the information for Picture 1.

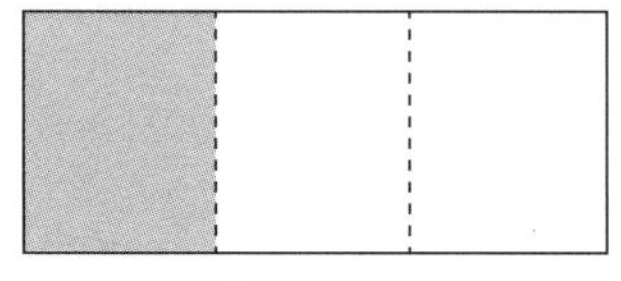

Picture 1

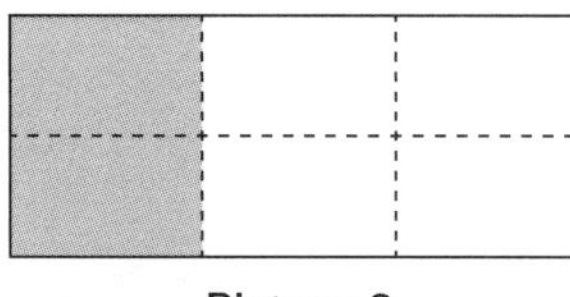

Picture 2

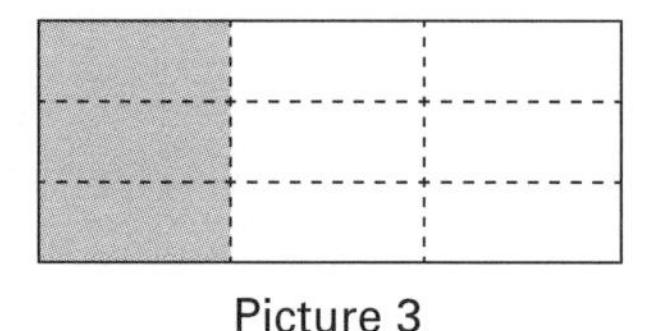

Picture 3

Picture number	Number of shaded parts	Total number of parts
1	*1*	*4*

2 Work with the class to carefully fold and shade Picture 2 shown above and record the information in the table. Then provide sheets of paper for volunteers to fold and shade 1 of 4 equal parts. Ask, *How can we fold the paper to get a greater number of equal parts? What is the total number of shaded equal parts that we can have? How do you know?* Invite students to make predictions, share their thinking, and then fold the paper to check. Fold and shade Picture 3 shown above and record the information in the table. Ask, *How does the picture number help you find the number of shaded parts?* (The number of shaded parts is the same as the picture number.) *How does the picture number help you find the total number of parts?* (The total number of parts is "double-double" the picture number.)

3 Ask the students to complete the blackline master.

Equal Pieces

Name ______________________________

1. Shade $\frac{1}{3}$ of each rectangle below. The first 2 have been done for you.

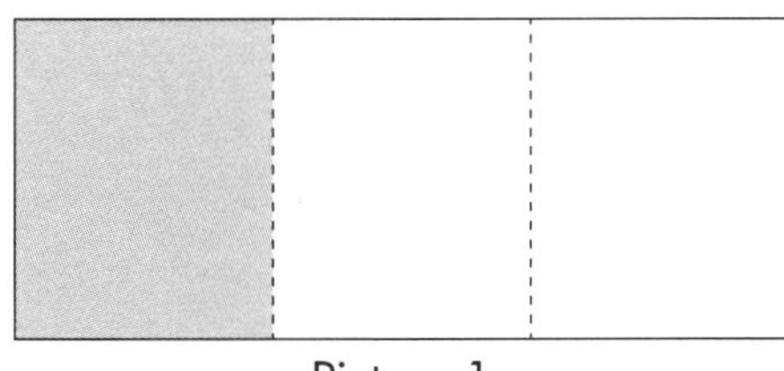

Picture 1

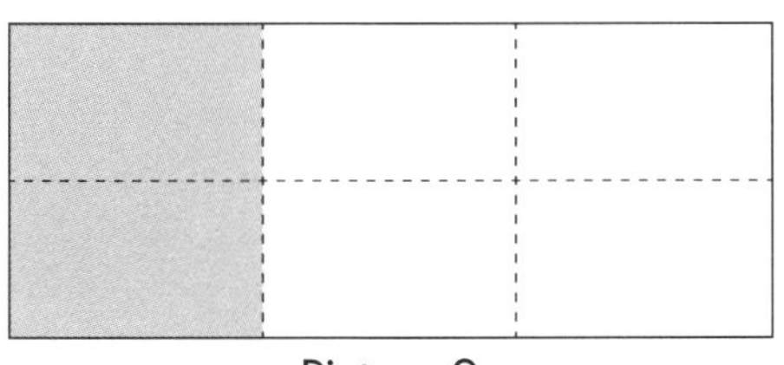

Picture 2

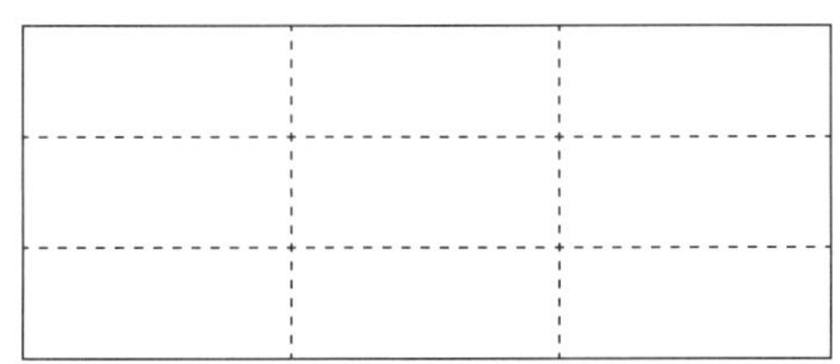

Picture 3

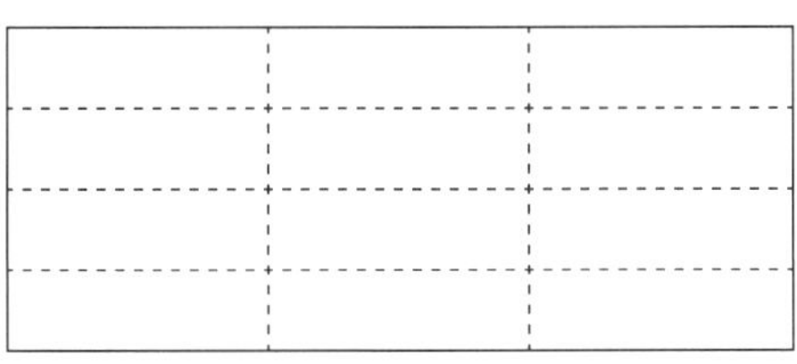

Picture 4

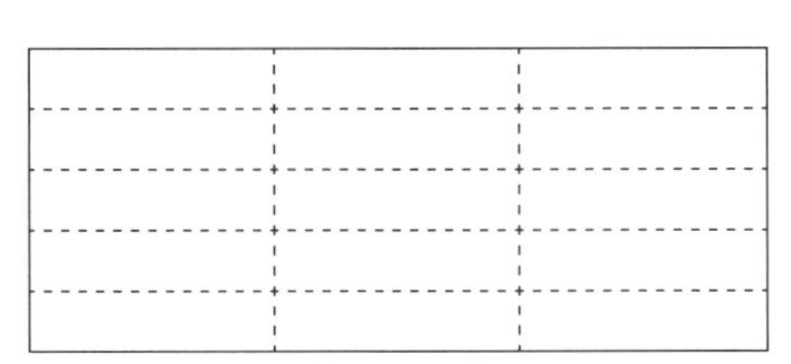

Picture 5

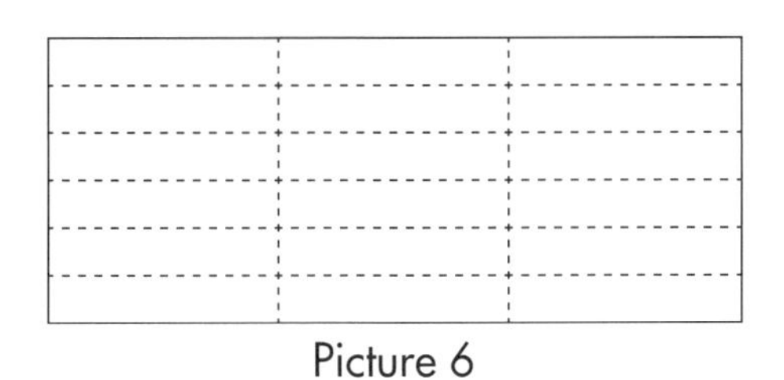

Picture 6

2. Complete this table.

Picture number	Number of parts shaded	Total number of parts
1	1	3
2	2	6
3		
4		
5		
6		

3. Write how you can use the picture number to figure out

   a. the number of parts shaded.

   ____

   ____

   b. the total number of parts.

   ____

   ____

4. Use your answers to Question 3 to complete this table.

Picture number	Number of parts shaded	Total number of parts
7		
10		
15		

Machine Pairs

Calculating output numbers for function machines in tandem

AIM

Students will calculate the output number when given an input number and addition rules for 2 function machines in tandem.

MATERIALS

- 1 copy of the blackline master (opposite) for each student

REFLECTION

Refer to Question 6 on the blackline master and ask, *How did you figure out the IN number for Machine 1?* (Subtract 12.) Discuss how we can figure out the IN number when given the OUT number, by using subtraction which is the inverse of addition.

1 Draw 2 function machines on the board, as shown below.

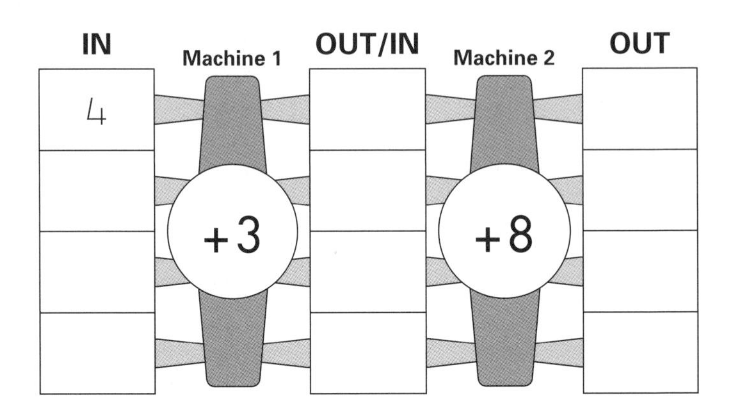

Say, *The OUT numbers for Machine 1 become the IN numbers for Machine 2. If 4 is the IN number for Machine 1, what will be the OUT number?* (7) *If 7 is the IN number for Machine 2, what will be the OUT number?* (15) Record the answers in the machines on the board. Repeat for the IN numbers 10, 3.5, and 28. Ask, *What is the total change for each IN number?* (+ 3 + 8 = + 11)

2 Ask the students to complete the blackline master. For each question, ask them to tell you one rule for both machines.

Machine Pairs

Name ______________________________

Complete each of these.

1.

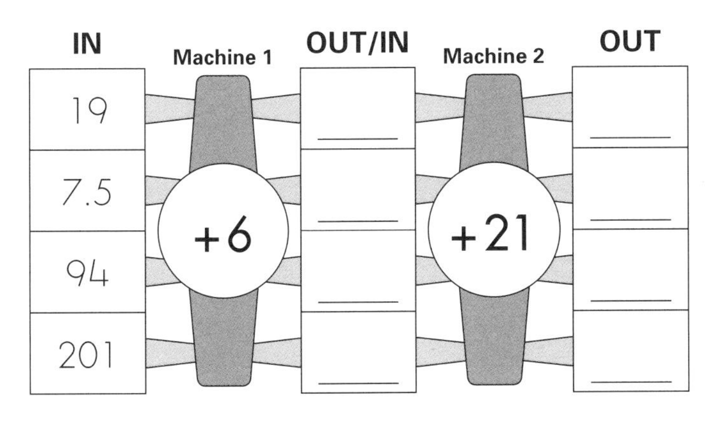

2.

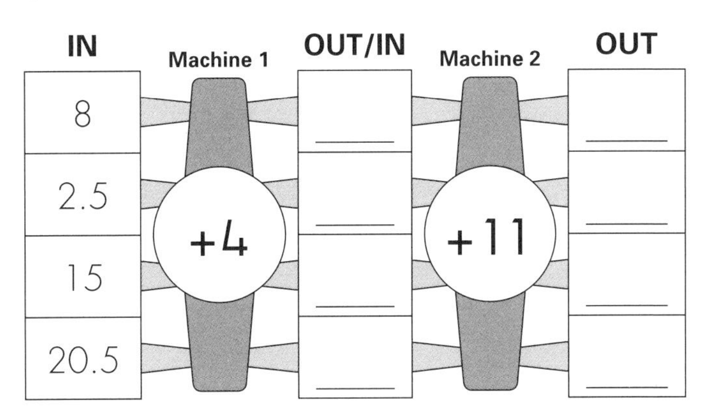

3.

IN	Machine 1	OUT/IN	Machine 2	OUT
71	+14	____	+8	____
15		____		____
117		____		____
21.5		____		____

4.

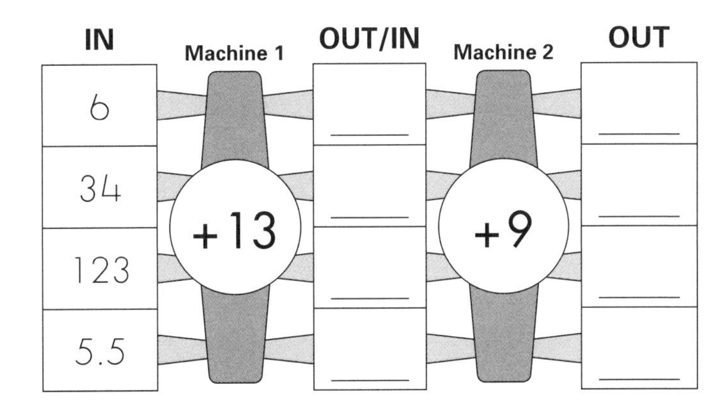

5.

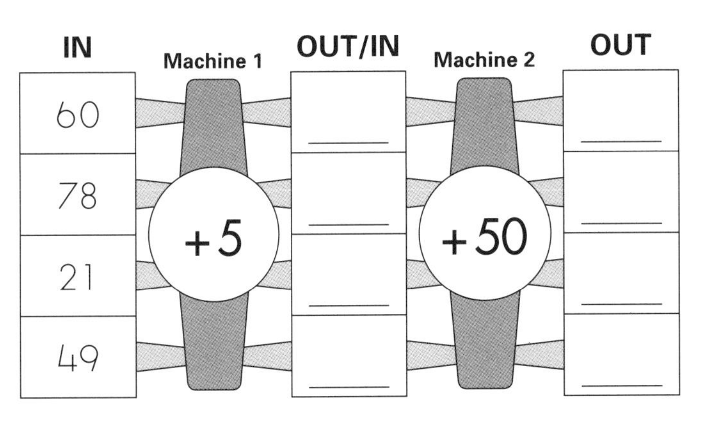

6.

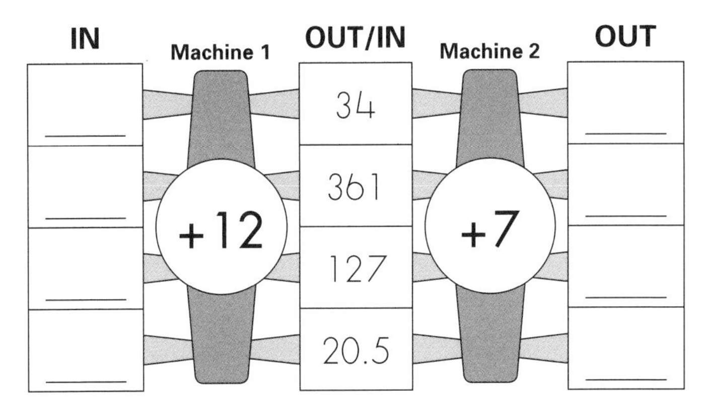

Patterns and Functions

7

Music Sale

Developing multiplication rules from functions

AIM

Students will calculate the output number when given the input number and the multiplication rule.

MATERIALS

- 1 copy of the blackline master (opposite) for each student

REFLECTION

Refer to Question 3a on the blackline master and ask, *How do we calculate the OUT number when we know the IN number? If the OUT number is 40, what will be the IN number? How do you know?*

1 Say, *CDs cost $2 each and DVDs cost $4 each. If we buy 3 CDs, how much will we spend? If we buy 10 CDs, how much will we spend?* Call on volunteers to suggest different numbers of CDs and ask other volunteers to figure out how much money they will spend. Ask the students to complete Question 1a on the blackline master. Call on volunteers to share their answers. Repeat the discussion for DVDs. Ask the students to complete Question 1b.

2 Have the students work in pairs to complete Question 2a. Direct one student to write the number of CDs, and the other student to calculate and write the total cost. They can then swap roles to complete Question 2b. Allow time for them to share their answers.

3 Draw a function machine on the board, as shown below.

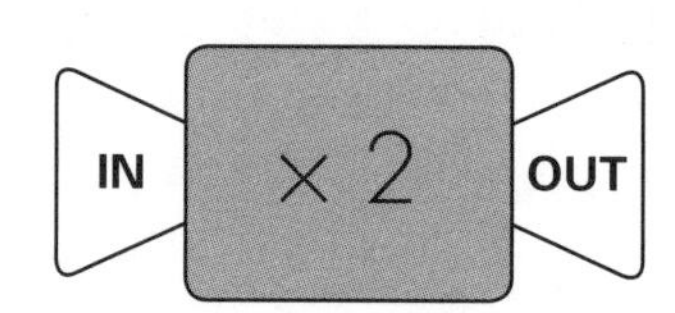

Ask, *If 6 goes into this machine, what number will come out?* Repeat for other IN numbers, such as 12, 21, 33, 42, 50, and 100. Ask the students to complete Question 3. Allow time for them to share their answers.

Music Sale

Name ______________________

1. Complete the tables.

a.

CDs cost $2 each	
Number of CDs	Total cost
2	
6	
12	

b.

DVDs cost $4 each	
Number of DVDs	Total cost
1	
3	
20	

2. Write some numbers of CDs and DVDs, then complete the tables.

a.

CDs cost $2 each	
Number of CDs	Total cost

b.

DVDs cost $4 each	
Number of DVDs	Total cost

3. Look at the rule. Write the OUT number.

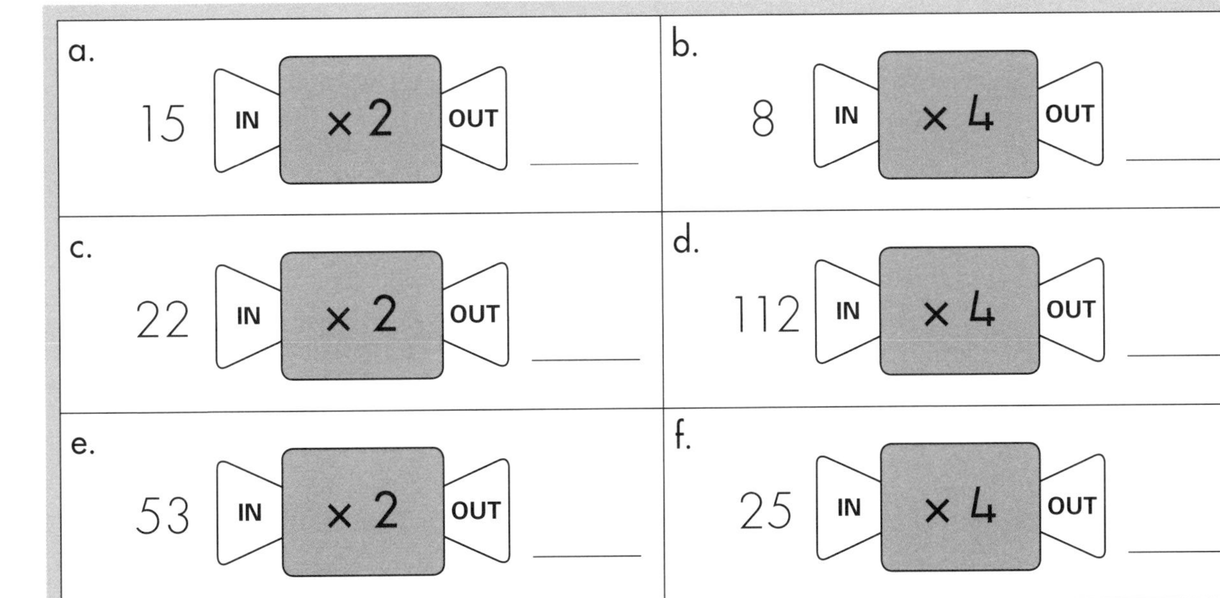

Patterns and Functions 8

CDs and DVDs

Using sharing as the inverse of multiplication

AIM

Students will calculate the input number when given the output number and the multiplication rule.

MATERIALS

- 1 copy of the blackline master (opposite) for each student
- Counters for each student

REFLECTION

Ask, *How do we calculate the OUT number when we know the IN number? If the OUT number is 40, and the rule is × 4, what will be the IN number? How do you know?* Discuss how the inverse of multiplication is division.

1 Say, *CDs cost $3 each and DVDs cost $5 each. If we spend $30 on CDs, how many do we buy? How do you know?* (Share 30 among 3.) Some students may need counters to model the division. Repeat for $21. Ask, *If we spend $25 on DVDs, how many do we buy? How do you know?* (Share 25 among 5.) *If we spend $55 on DVDs, how many do we buy?* Call on volunteers to share their solutions.

2 Ask the students to complete Question 1 on the blackline master. Call on volunteers to share their answers. Have the students work in pairs to complete Question 2a. Direct one student to write the total cost of the CDs, and the other student to figure out the number of CDs. They can then swap roles to complete Question 2b. Allow time for them to share and justify their answers.

3 Draw a function machine on the board, as shown below.

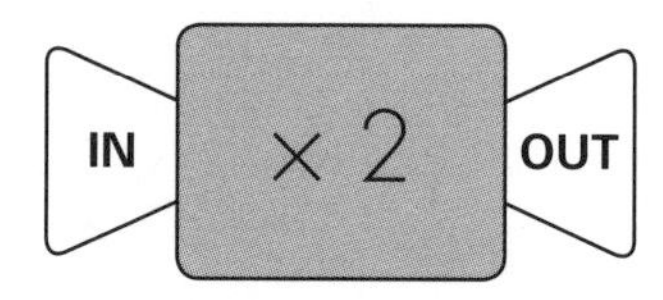

Ask, *If 50 comes out of this machine, what was the IN number?* Discuss how the students solve the problem. Repeat for 24, 18, and 48.

4 Ask the students to complete the blackline master. Call on volunteers to share their IN numbers.

CDs and DVDs

Name ____________________

1. Complete the tables.

a.

CDs cost $3 each	
Number of CDs	Total cost
	$15
	$27
	$48

b.

DVDs cost $5 each	
Number of DVDs	Total cost
	$30
	$45
	$105

2. Write some total costs of CDs and DVDs, then complete the tables.

a.

CDs cost $3 each	
Number of CDs	Total cost

b.

DVDs cost $5 each	
Number of DVDs	Total cost

3. Look at the rule. Write the IN number.

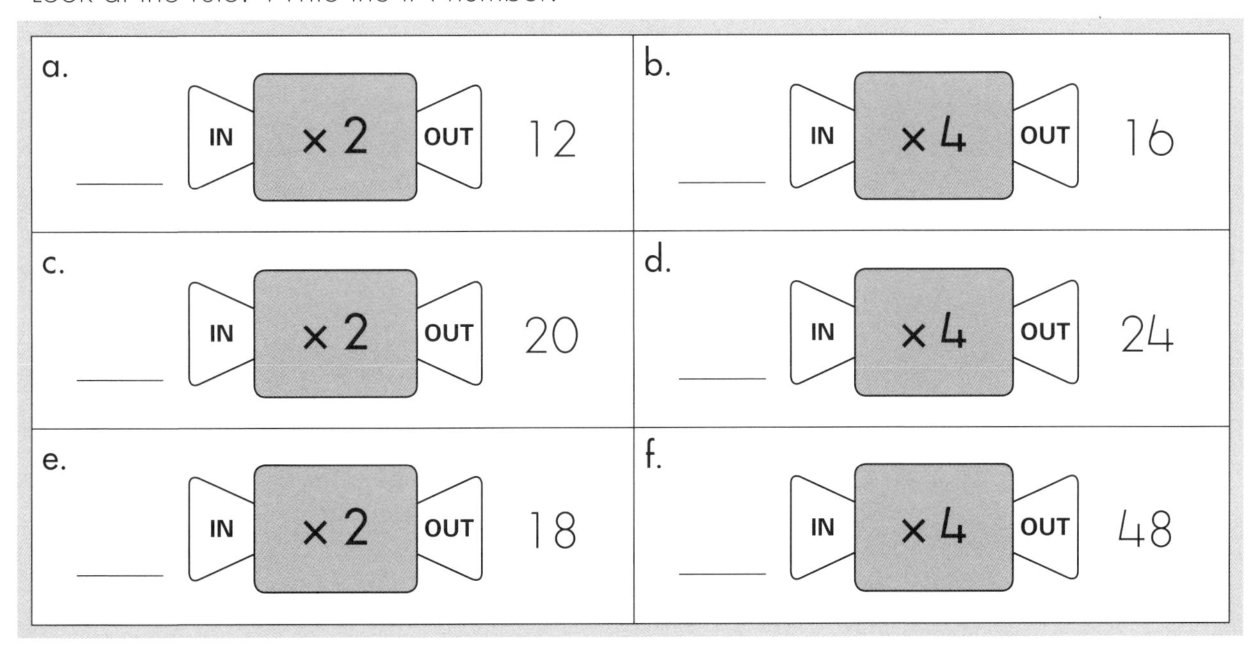

Magic Machines

Figuring out rules for multiplication, addition, and subtraction situations

AIM

Students will figure out the rules for multiplication, addition, and subtraction function machines.

MATERIALS

- 1 copy of the blackline master (opposite) for each student

REFLECTION

Ask, *What operation is the opposite of addition? What operation is the opposite of multiplication?* Discuss inverse operations.

1 On the board, draw a function machine as shown below.

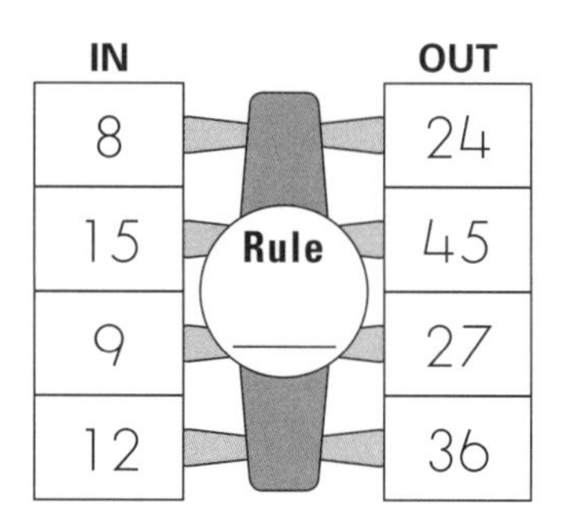

Say, *This is a function machine.* Point to the numbers at the left and say, *These are the IN numbers and on the other side are the OUT numbers. What is the function rule for this machine?* (x 3) Write the rule on the machine and ask, *How did you figure it out?* (Multiply by 3.) Erase the rule and change the OUT numbers to 16, 23, 17, and 20. Ask, *What is the function rule for these numbers?* (+ 8) *How did you figure it out?* (Add 8.)

2 On the board, draw 2 function machines as shown below.

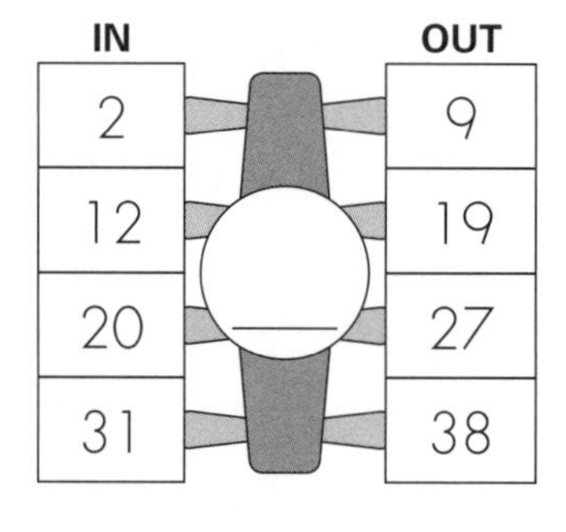

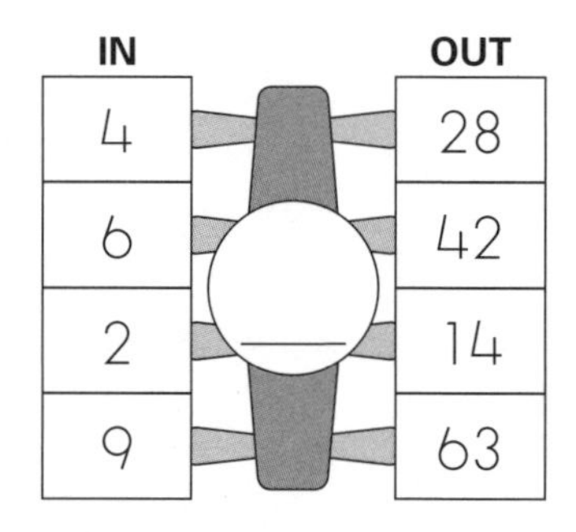

Point to the 1st machine and say, *This function machine changes these IN numbers to these OUT numbers.* Point to the 2nd machine and say, *This function machine changes these IN numbers to these OUT numbers. What is the rule for each of these function machines?* (1st machine is + 7, 2nd machine is x 7.) Discuss the difference between adding 7 and multiplying by 7.

3 Ask the students to complete the blackline master. Call on volunteers to share their rules.

Magic Machines

Name ____________________

Write a rule for each machine, then write the missing number.

1. 2.

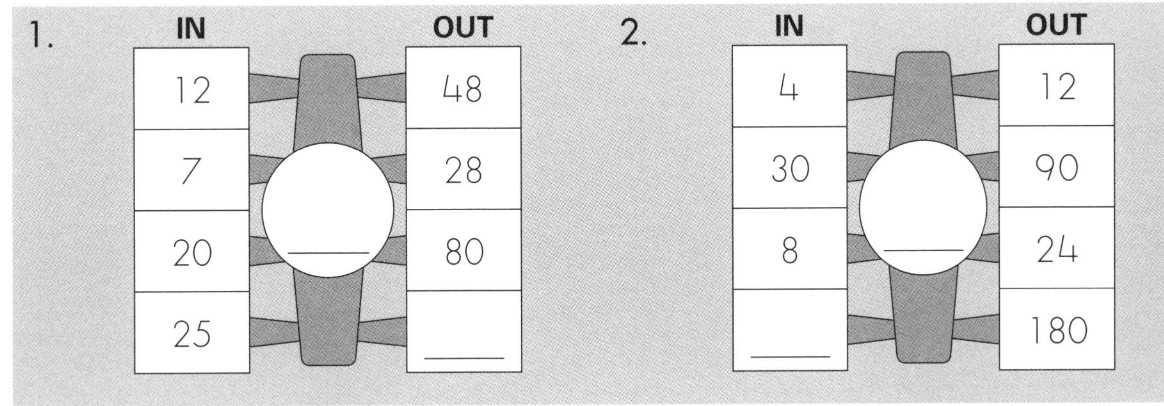

3.

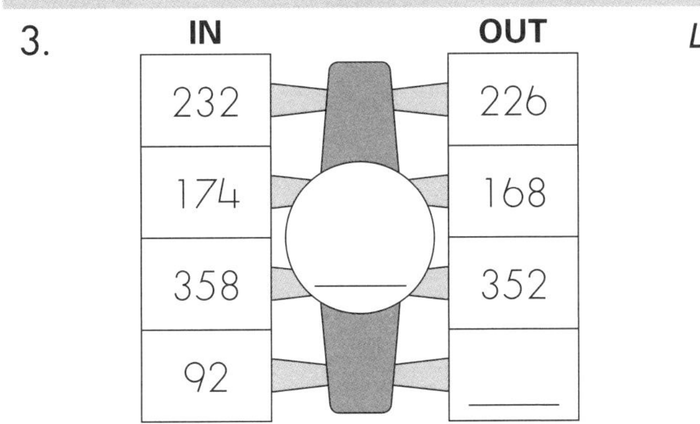

4.

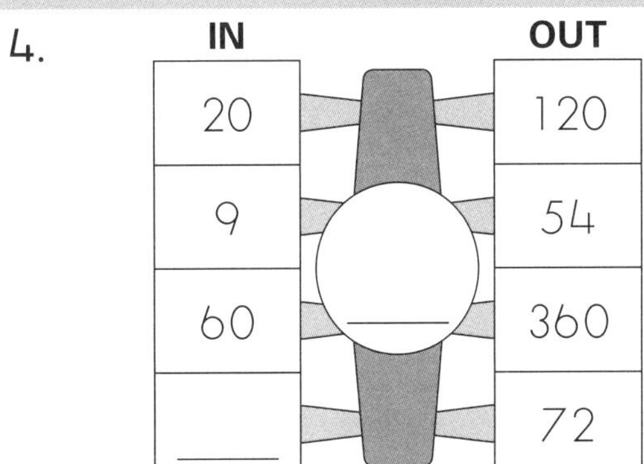

5. 6.

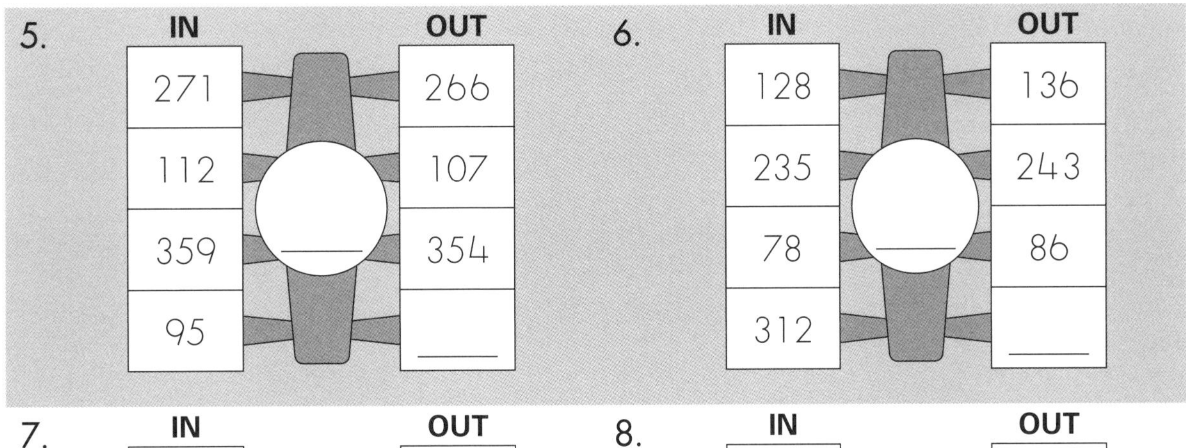

7.

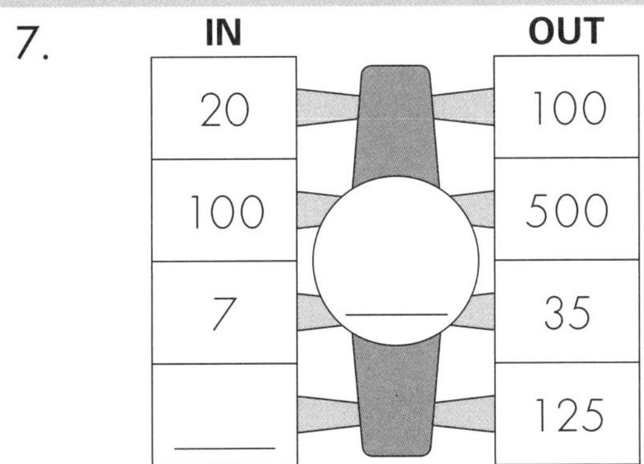

8.

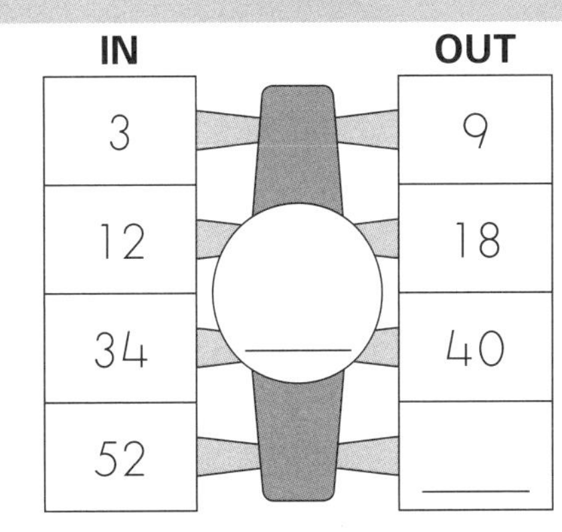

Double Rules

Figuring out rules for addition and subtraction machines

AIM

Students will figure out rules for function machines in tandem, find a range of rules, and discuss "equivalence".

MATERIALS

- 1 copy of the blackline master (opposite) for each student

REFLECTION

Challenge the students to draw another 2 linked function machines and a list of IN numbers and matching OUT numbers. Direct them to swap machines with another student and figure out the possible rules.

1 On the board, draw the function machines shown below.

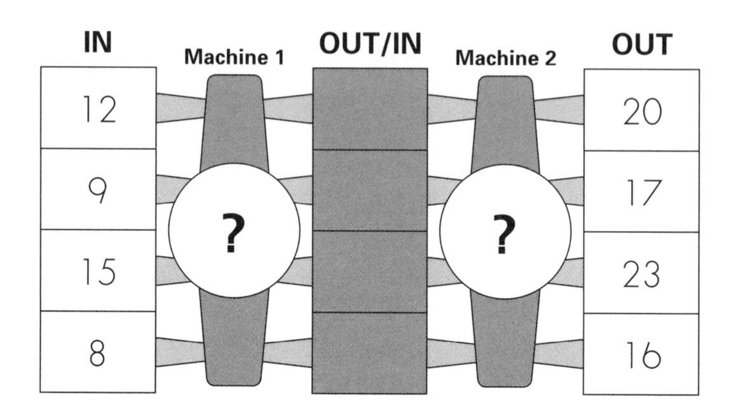

Say, *These function machines change IN numbers into OUT numbers. Each machine has its own rule. Both rules use the same operation. What is the total change for these 2 rules?* (+ 8) *What is the rule for each function machine? How did you figure it out?* Call on volunteers to share their ideas. Ask, *How many pairs of rules are there?* (+ 8, + 0; + 7, + 1; + 6, + 2; + 5, + 3; + 4, + 4) Write the ordered list on the board. Repeat for 2 sets of numbers with a total change of – 9.

2 Ask the students to complete Question 1 on the blackline master. Call on volunteers to share their solutions. Discuss how they figured out all the possible rules.

Double Rules

Name ______________________________

1. Complete the table to show some possible rules for these machines. Both operations must be the same.

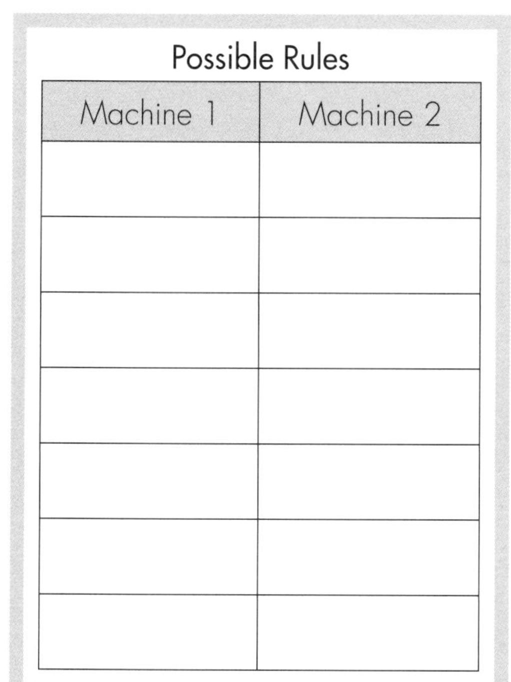

Possible Rules

Machine 1	Machine 2

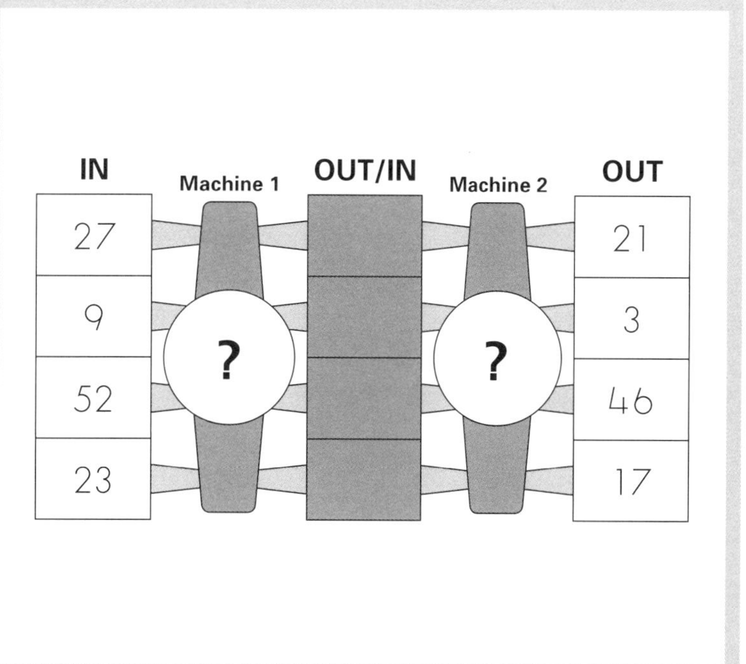

2. Complete this table. Both operations must be the same.

Possible Rules

Machine 1	Machine 2

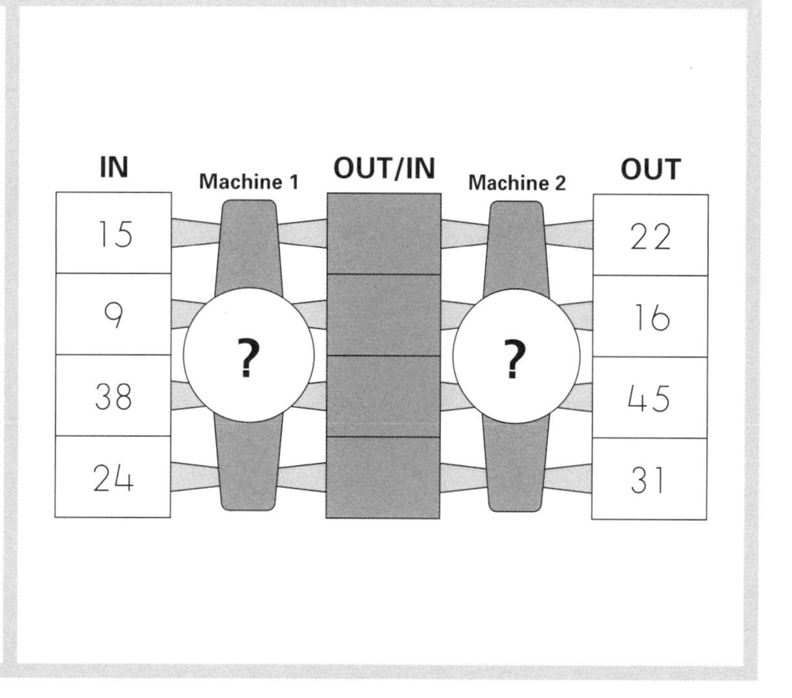

Great Groups

Turnarounds for multiplication as repeated addition

AIM

Students will conclude that when the order of multiplication is changed in repeated addition situations, the answer stays the same.

MATERIALS

- Magnetic counters (or ordinary counters and Blu-Tack)
- 1 copy of the blackline master (opposite) for each student

REFLECTION

Ask, *Is 8 groups of 3 the same as 3 groups of 8? How do you know?* Call on volunteers to share their thinking. Ask, *If we change the order in which we multiply 2 numbers, will the answer be the same or different? Does it matter how big the 2 numbers are?* Call on volunteers to share their thinking.

1 Place 3 groups of 2 counters on the board. Ask, *How many groups are there? How many counters are in each group? How many counters are there in total? How do we say this?* (3 groups of 2 equals 6.) Move the last 2 counters to show 2 groups of 3 counters. Ask, *How many groups are there now? How many counters are in each group? How many counters are there in total? How do we say this?* (2 groups of 3 equals 6.) *Are 2 groups of 3 the same as 3 groups of 2?* Write ***2 groups of 3 = 3 groups of 2*** and ***2 x 3 = 3 x 2*** on the board. Repeat for 2 x 5 and 5 x 2, and 4 x 5 and 5 x 4. Each time, move the counters to show the change. Explain that these equations are called "turnarounds" because we can turn around the order of the parts and still have the same total amount.

2 Ask the students to complete the blackline master. Call on volunteers to share and explain their answers.

Great Groups

Name ______________________________

1. Complete these equations.

a.

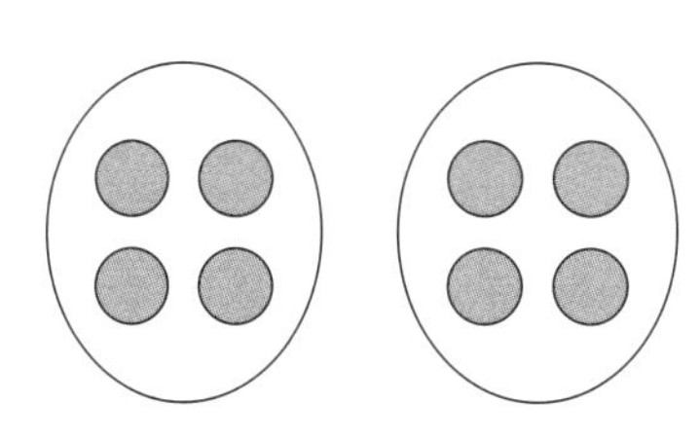

2 groups of ______ = 8

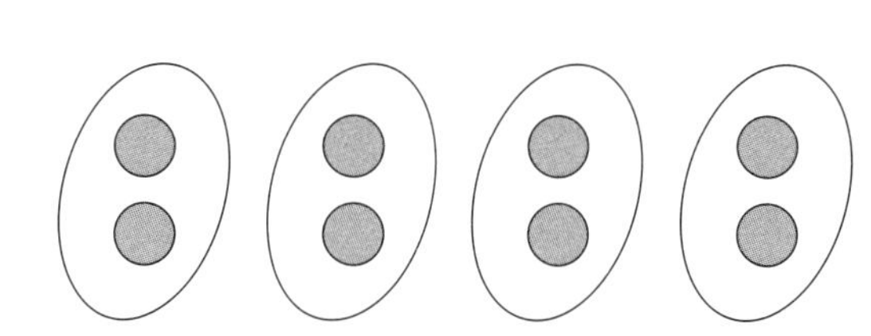

4 groups of ______ = ______

2 groups of ______ = 4 groups of ______

b.

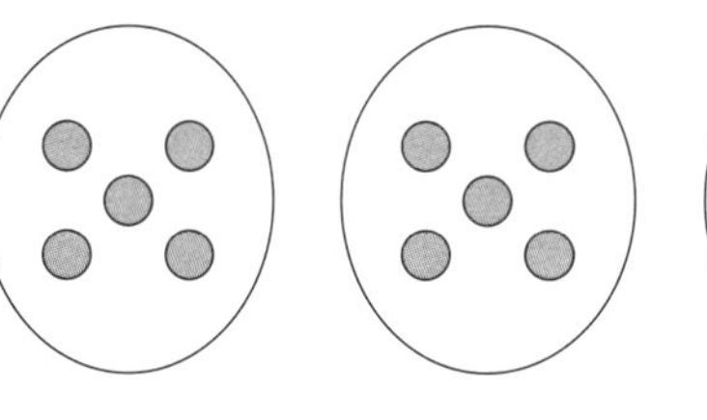

______ groups of 5 = ______

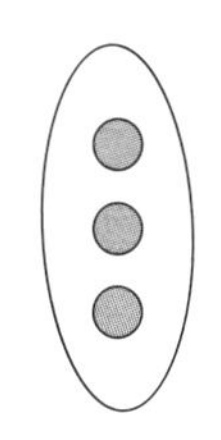
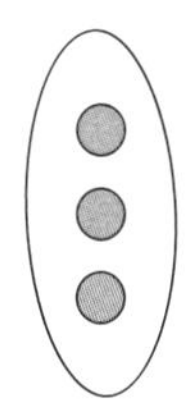
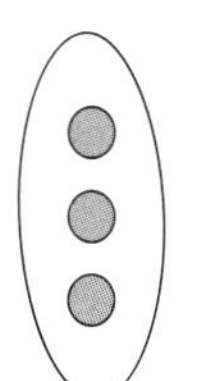
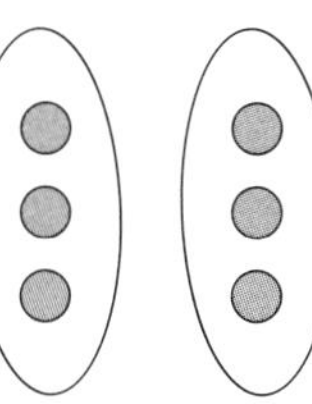

5 groups of ______ = ______

______ groups of ______ = ______ groups of ______

2. Draw counters to show these groups, then complete the equation.

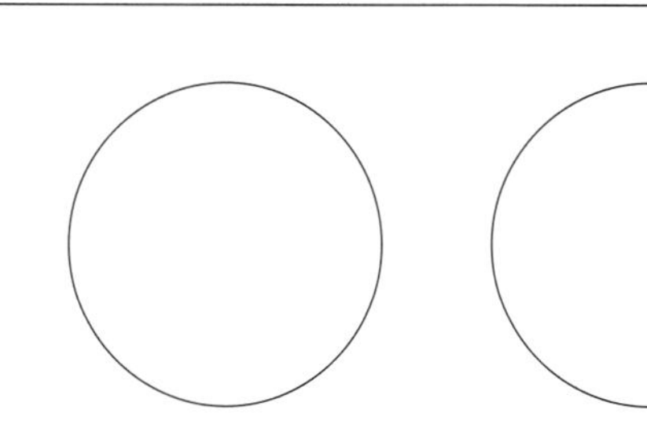

3 in each group

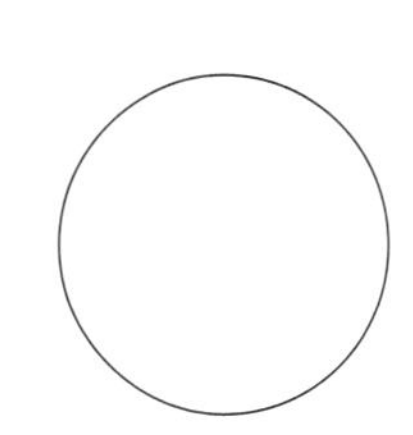

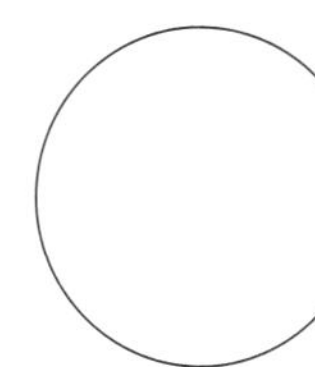

2 in each group

______ groups of ______ = ______ groups of ______

Properties | 1

Hot and Cold

Changing the order in which two numbers are subtracted

AIM

Students will use a number line to explore the idea of turnarounds for subtraction and conclude that turnarounds are not possible for subtraction.

MATERIALS

- Large number track for –20 to +20
- 1 copy of the blackline master (opposite) for each student

REFLECTION

Ask, *What happens when we subtract the same pair of numbers in a different order? Do we get the same answer?* Call on volunteers to share their ideas. Discuss how we can change the order in which we add 2 numbers but not the order in which we subtract 2 numbers.

1 Say, *Imagine you are living in a place where it is very cold in winter. If the temperature is 8 degrees and it drops by 5 degrees, what is the new temperature? How do you know?* Invite volunteers to suggest places where it might get this cold in winter. Write ***8 – 5 = 3*** on the board. Invite a volunteer to stand on the number track at zero. Ask, *What was the temperature at the start?* Instruct the student to walk to 8. Ask, *By how much did the temperature drop? In which direction should (Zac) walk? How many steps should (he) take?* Instruct the student to turn and walk back 5 spaces. Repeat for ***11 – 6 = 6*** and ***16 – 9 = 7*** with different students walking the number track.

2 Say, *Imagine the temperature is 5 degrees and it drops by 8 degrees. What is the new temperature? Is this temperature the same as starting at 8 degrees and dropping by 5 degrees? Are the answers the same? How can we use the number track to show that they are different?* Call on a volunteer to walk from 5 to 8 and another to walk from 8 to 5. Ask, *Is 5 subtract 8 the same as 8 subtract 5?* Draw attention to the positions of the students on the number track at +3 and –3. Ask, *How can we write this?* Write ***5 – 8 = –3*** and ***8 – 5 = 3*** on the board. Discuss how the answers are not equal, so we cannot write turnarounds for subtraction.

3 Ask the students to complete the blackline master. Call on volunteers to share their answers.

Hot and Cold

Name ______________________

1. Draw jumps on the thermometer to help you complete these equations.

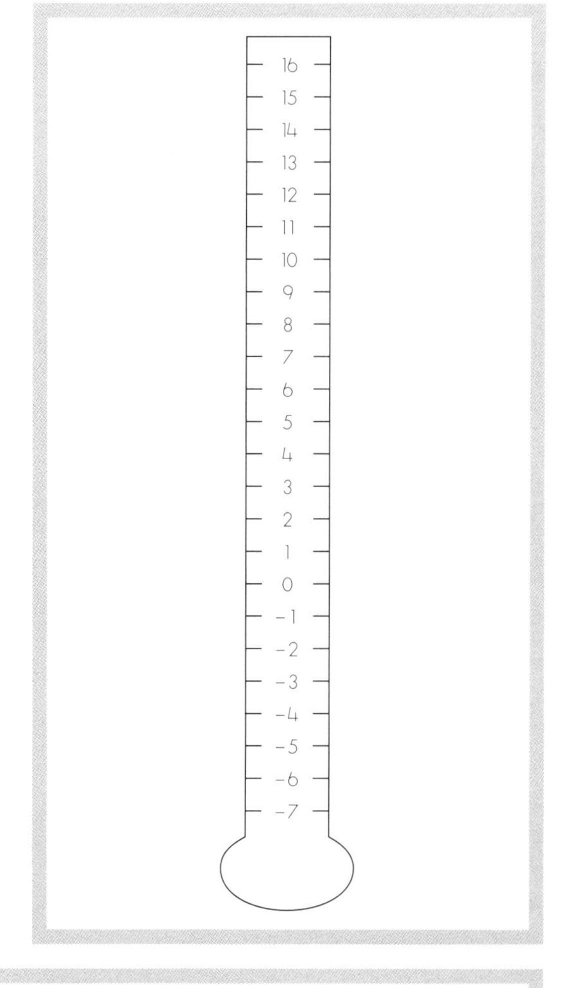

a. 9 degrees – 6 degrees = ______ degrees

6 degrees – 9 degrees = ______ degrees

b. 7 degrees + 8 degrees = ______ degrees

8 degrees + 7 degrees = ______ degrees

c. 12 – 5 = ______

5 – 12 = ______

d. 4 + 10 = ______

10 + 4 = ______

2. Write = or ≠ to complete these number sentences. Draw jumps to help you.

a.

3 – 16 ____ 16 – 3

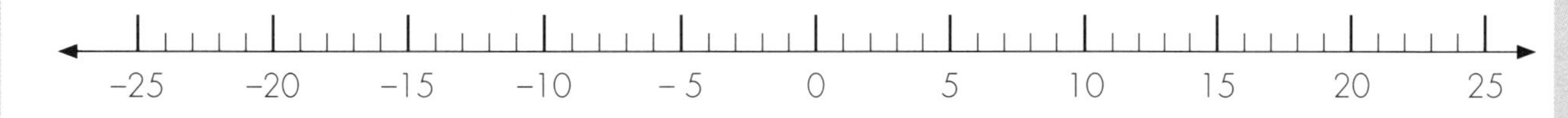

b.

3 + 12 ____ 12 + 3

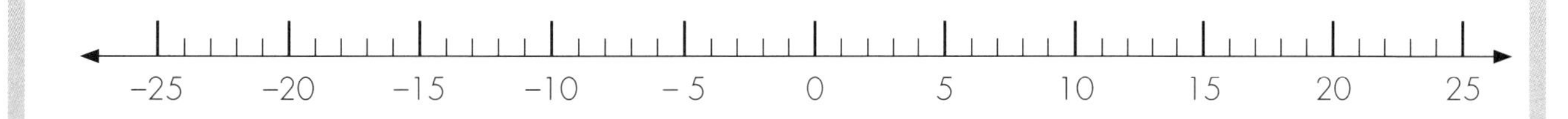

Properties 2

Mix and Match

Exploring turnarounds for multiplication

AIM

Students will use tree diagrams to show that if the order of multiplication is changed the answer remains the same.

MATERIALS

- 1 copy of the blackline master (opposite) for each student

REFLECTION

Refer to the blackline master. Identify individuals who started their tree diagram with drinks and those who started their diagram with fruit. Invite one student to draw each type of diagram on the board. Check that the number of combinations is the same for both diagrams and write the 2 number sentences. Make sure the students know that the order is not important.

1 On the board, draw 2 different pairs of jeans and 3 different shirts. Ask, *How many different outfits can we make? How can we show all the possibilities?* Call on volunteers to share their solutions, then make an ordered list on the board and conclude that there are 6 different outfits. Say, *We can also use a tree diagram to show all the different outfits.* Work with the class to draw the tree diagrams shown below on the board.

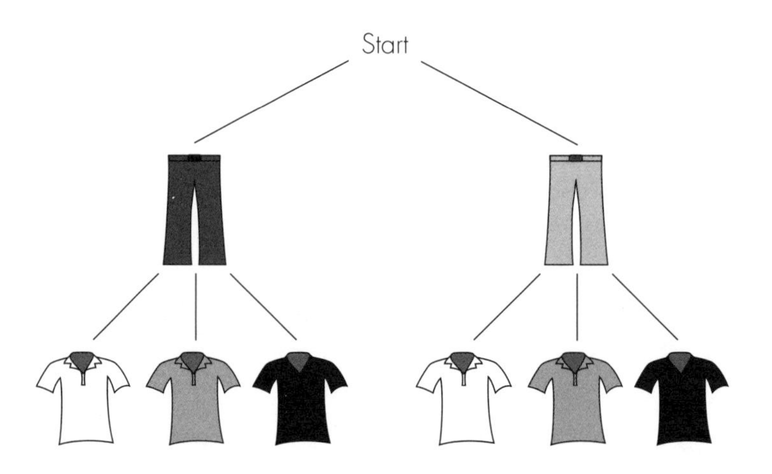

Discuss how the diagrams show 2 groups of 3 outfits, and we can figure out the number of different outfits by multiplying the number of groups by the number of outfits. On the board, write ***2 x 3 = 6***.

2 Repeat the discussion, starting with the shirts and then selecting the jeans. Write the equation ***3 x 2 = 6*** on the board. Ask, *Do we have the same outfits as we had before? Is the number of different outfits the same each time? How do we write this?* Write ***2 x 3 = 3 x 2*** on the board. Repeat Steps 1 and 2 for ***5 x 3 = 3 x 5***.

3 Ask the students to complete the blackline master.

Mix and Match

Name ___________________________

A snack includes a drink and a piece of fruit. Draw a tree diagram to show the number of different snacks.

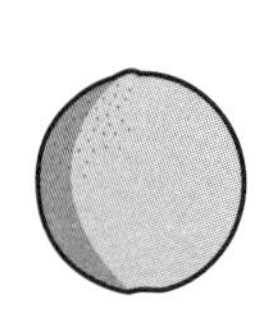

Number of different snacks = ______

Properties 3

Getting Older

Increasing both parts of a subtraction problem by the same amount

AIM

Students will explore the part–part–total pattern for subtraction equations; that is, if one part of a subtraction equation is increased, in order to obtain the same answer, the other part must be increased by the same amount.

MATERIALS

- Connecting cubes
- 1 copy of the blackline master (opposite) for each student

REFLECTION

Refer to the blackline master and ask the students to describe what they notice about the numbers in the columns of the table. Reinforce the idea that the difference stays the same when the minuend and subtrahend of a subtraction sentence are both increased/decreased by the same amount.

TEACHING NOTE

If the students are not confident with the terms minuend and subtrahend, refer to each part of the subtraction expression as simply the 1st number and 2nd number.

1 Say, *Sebastian and Laura are brother and sister. Sebastian is 12 years old and Laura is 8 years old. What is the difference between their ages? How do you know? What number sentence can we write to show this?* Invite students to model the ages with cubes, then write ***12 – 8 = 4*** on the board. Ask, *How old will Sebastian be in 1 year? How old will Laura be in 1 year? How does each age change?* (Each age increases by 1.) *How can we write this as a number sentence?* Write ***13 – 9 = 4*** or ***(12 + 1) – (8 + 1) = 4*** on the board. Ask, *Does the difference between their ages remain the same?* (Yes.)

2 Write ***12 – 8 =*** ___ on the board and draw a number line and arrow to represent the equation, as shown below.

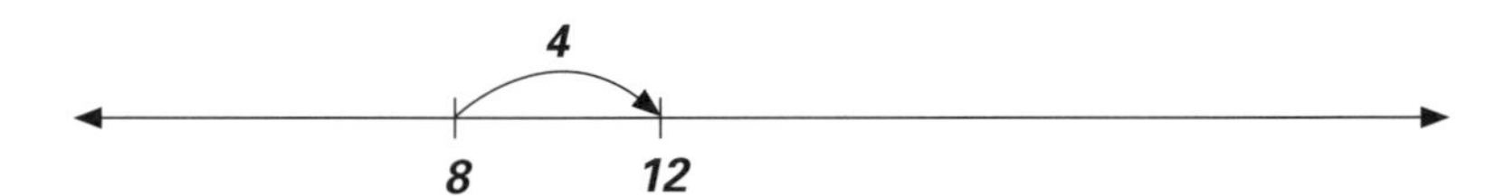

Ask, *What is the difference between their ages? How can we show the difference between their ages for next year?* Ask a volunteer to draw an arrow on the number line and write the related number sentence.

Repeat for the difference between Sebastian's and Laura's ages in 3 years and then in 5 years.

3 Read the blackline master with the class. Allow time for the students to complete the questions.

Getting Older

Name ____________________

Mick is now 12 years old and Carl is now 9 years old.

1. Draw a jump to show the difference between their ages.

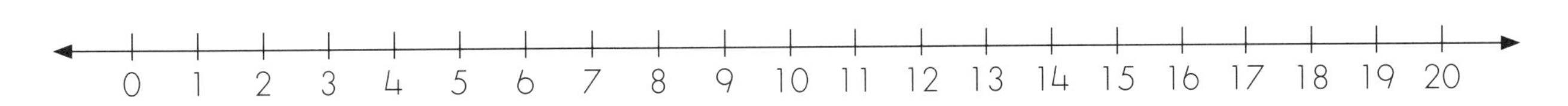

2. Complete the table. Draw jumps to help you.

	Mick's age	Carl's age	Difference
1 year ago			______ = ☐
Now	12	9	$12 - 9 =$ ☐
1 year from now			______ = ☐
2 years from now			______ = ☐

3. Look at the numbers in the shaded columns. What do you notice? ____________________

____

4. What do you notice about the differences? ____________________

____

5. Write an equation to show the difference between their ages

a. 5 years from now. ____________________

b. 10 years from now. ____________________

c. 15 years from now. ____________________

Properties 4

Make it Easy

Investigating the part–part–total relationship for addition

AIM

Students will explore changes that can be made to the addends of an addition sentence so that the total remains the same.

MATERIALS

- 1 copy of the blackline master (opposite) for each student

REFLECTION

Refer to Question 2 on the blackline master and ask several students to describe how they changed the numbers to make the calculations easier. Emphasize that when one addend is decreased, the other addend is increased by the same amount so that the total remains the same.

1 On the board, draw a number line as shown below.

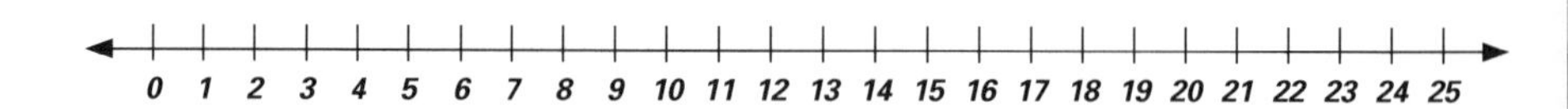

Say, *If we start at 0 and make 2 jumps to reach 21, what could be the lengths of the 2 jumps? How do you know?* Invite volunteers to draw the jumps and record the related number sentences on the board. Arrange the number sentences in the order shown.

14 + 7 = 21
15 + 6 = 21
16 + 5 = 21

On the board, write a sentence with an unknown, such as __ ***+ 4 = 21*** to extend the pattern above. Ask, *What is the value of the unknown?* (17) Encourage students to explain how we can make one jump longer and shorten the other jump at the same time. Discuss how this means that in a number sentence we increase one addend by the same amount that we decrease the other addend. If time allows, repeat for 2 jumps to reach 24.

2 On the board, draw the price tags shown below.

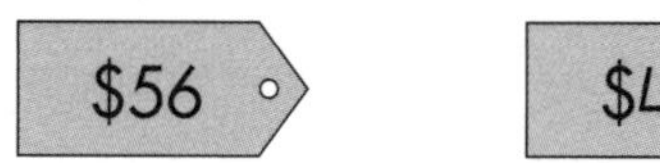

Ask, *What number sentence can we write to figure out the total cost?* On the board, write ***56 + 49 = 105***. Ask, *How can we rewrite this to show another way to figure out the total cost?* For example, write ***55 + 50 = 105*** on the board. Invite volunteers to describe their thinking. Emphasize strategies that increase one addend and decrease the other addend. Repeat the discussion for other pairs of prices, such as $46 and $38.

3 Ask the students to complete the blackline master. Allow time for them to share their answers.

Make it Easy

Name ______________________

1. Write price tags to show different ways to spend $20 on 2 items.

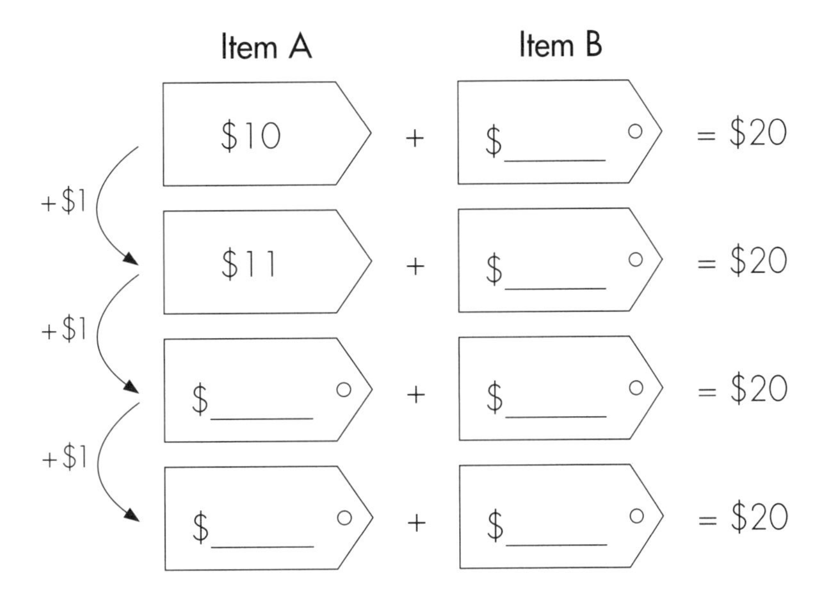

a. Look at how the price of Item A changes. What do you notice?

____

b. Look at how the price of Item B changes. What do you notice?

____

c. How do the 2 prices change so that the total stays the same?

____

2. Write new numbers that make it easier to figure out the total cost.

a. ______ + ______ = $______ = $49 + $26

b. ______ + ______ = $______ = $37 + $58

c. ______ + ______ = $______ = $43 + $59

d. ______ + ______ = $______ = $28 + $44

Properties 5

Grids Galore

Using a coordinate grid to explore relationships

AIM

Students will use a coordinate grid to explore relationships.

MATERIALS

- 1 copy of the blackline master (opposite) for each student

REFLECTION

Refer to the grid on the board and ask, *If someone has the same number of sisters as brothers, where will they be on the grid?* (On the diagonal.) *If someone has the greatest number of sisters, where will all the other spaces on the grid be?* (Below this space in the vertical direction.) *If someone has the fewest number of brothers, where will all the other spaces be?* (To the right of this space on the horizontal axis.)

1 On the board, draw the grid and write the clues, as shown below.

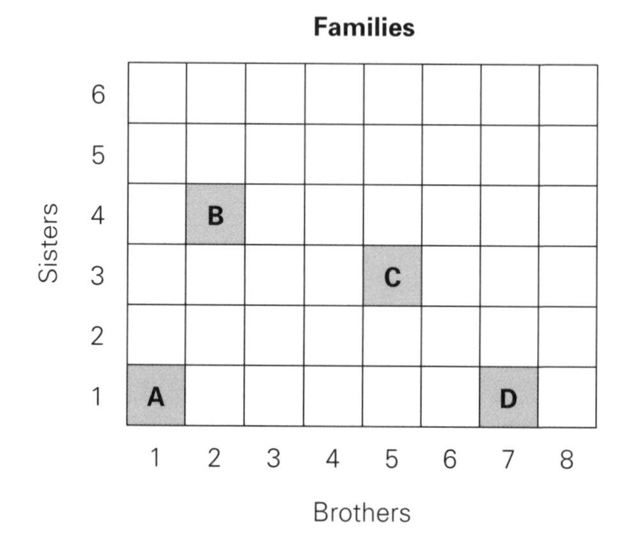

Clues

- ***Ben has more brothers than Lucy.***
- ***David has the fewest number of brothers.***
- ***Mei has 1 more sister than Lucy.***

Ask, *What do the shaded spaces on the grid represent? How do we decide which space represents each person? How many brothers and sisters does A have? Which space is David? How did you figure it out?* (A is the space closest to the vertical axis — the fewest brothers.) *Which spaces show that a person has 1 more sister than another person?* (B and C.) *Which space is Mei? Which space is Lucy? Why?* (B is Mei and C is Lucy because Mei has 1 more sister than Lucy.) *Which space is Ben?* (Ben must be D.)

2 Ask the students to complete the blackline master. Call on volunteers to share and justify their answers.

Grids Galore

Name ______________________________

1. Look at the grid, read the clues, then answer the questions.

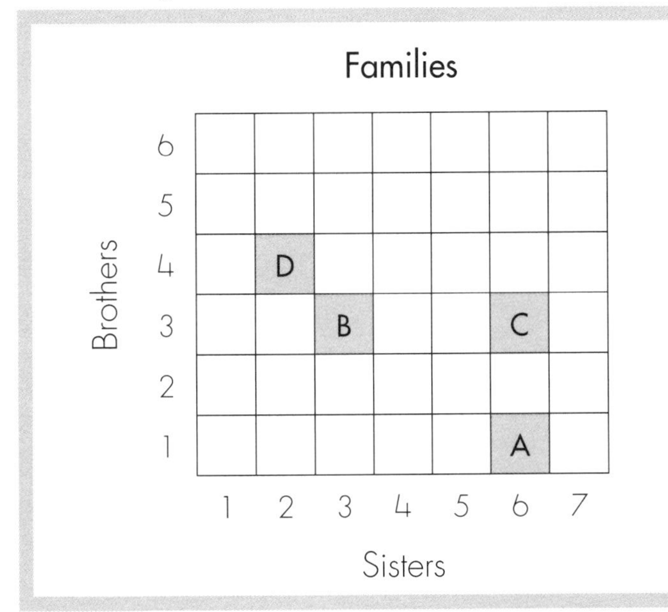

CLUES

Jesse and Emma have the same number of brothers.

Marla and Jesse have the same number of sisters.

Marla has more sisters than Emma.

Ralph has the greatest number of brothers.

a. Who is A? ____________________ b. Who is B? ____________________

c. Who is C? ____________________ d. Who is D? ____________________

2. Look at the grid, read the clues, then answer the questions.

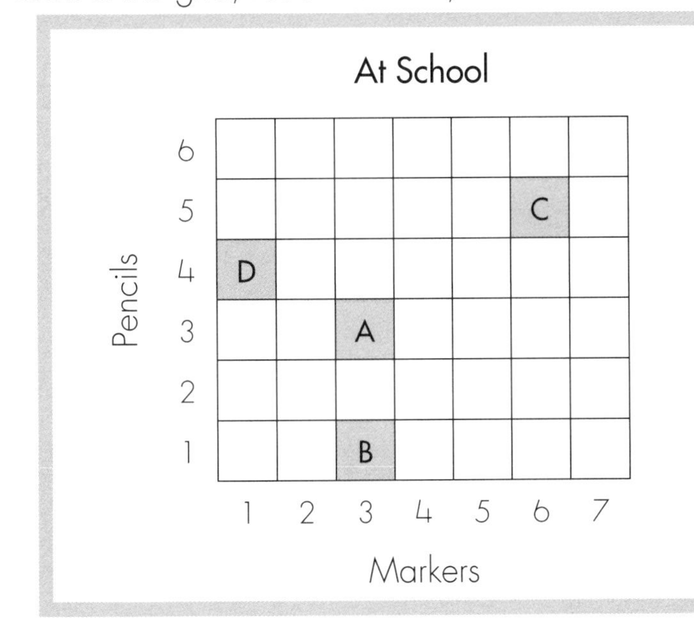

CLUES

Anna has the same number of markers as pencils.

Thomas has one more pencil than Anna and one less than Sally.

Mark has the fewest number of pencils.

a. Who is A? ____________________ b. Who is B? ____________________

c. Who is C? ____________________ d. Who is D? ____________________

Representations 1

Home, Sweet Home

Using a coordinate grid to explore relationships

AIM

Students will use a coordinate grid to explore relationships.

MATERIALS

- 1 copy of the blackline master (opposite) for each student

REFLECTION

Refer to the table on the board and ask, *If someone has as many family members as telephones, where will they be on the grid?* (On the diagonal.) Discuss how Simon, with the most telephones, is above all the other spaces, and Sergio, with the most family members, is the space at the greatest distance to the right.

1 On the board, draw the grid and write the clues, as shown below.

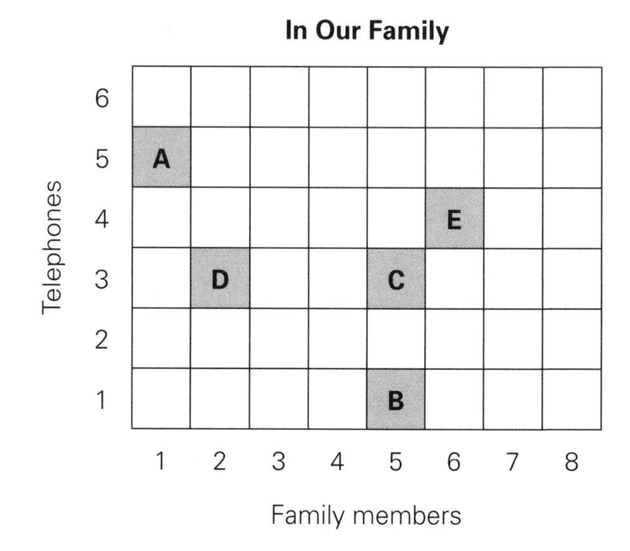

Clues

- ***Justin and Madison have the same number of telephones.***
- ***Phoebe and Justin have the same number of family members.***
- ***Sergio has 1 more family member and 1 more telephone than Justin.***
- ***Simon has the most telephones.***

Ask, *What do the shaded spaces on the grid represent? How do we decide which space represents each person? How many family members and telephones does A have? Which space is Justin?* (C.) *Why?* (He has the same number of family members as D, and the same number of telephones as B.) *Which space is Simon?* (A. His family has the most telephones.) *Which space is Sergio?* (E.) *Which space is Madison?* (D. She has the same number of telephones as Justin.) *Which space is Phoebe?* (B. She has the same number of family members as Justin.)

2 Ask the students to complete the blackline master. Call on volunteers to share and justify their answers.

Home, Sweet Home

Name ______________________

1. Look at the grid, read the clues, then answer the questions.

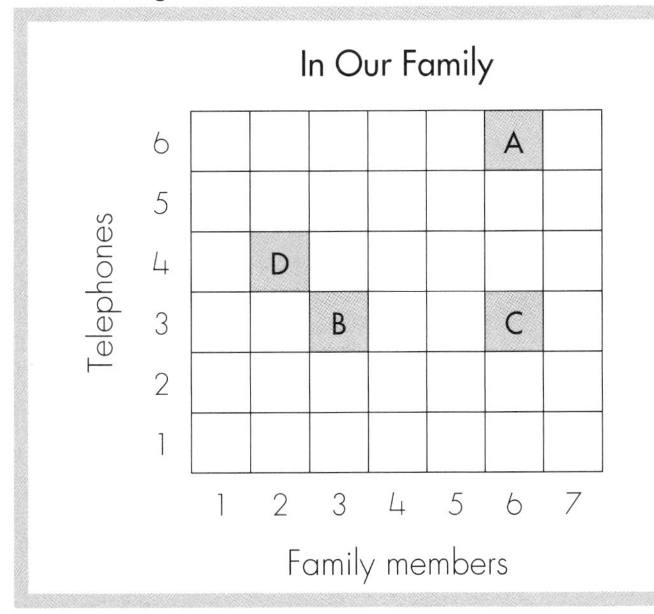

CLUES

Maria has the greatest number of telephones.

Theo has as many telephones as family members.

Ryan has half as many family members as telephones.

Theo and Carole have the same number of telephones.

a. Who is A? ______________ b. Who is B? ______________

c. Who is C? ______________ d. Who is D? ______________

2. Look at the grid, read the clues, then answer the questions.

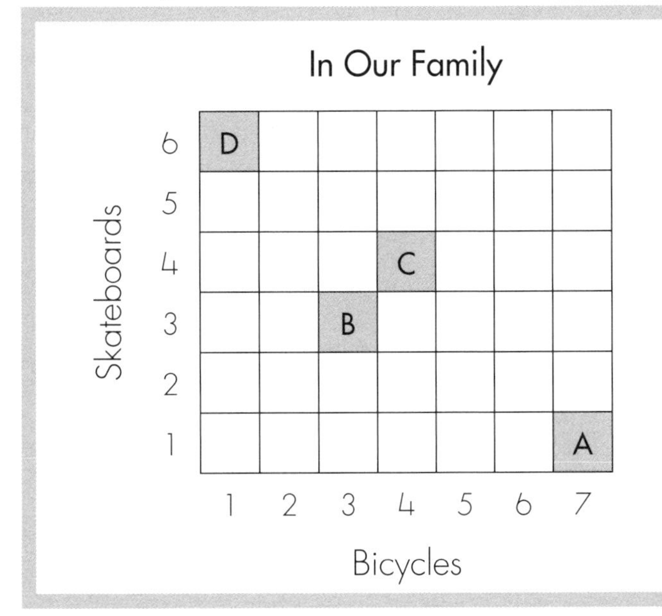

CLUES

Aaron's family has the greatest number of skateboards.

Mike's family has 7 bicycles.

Tyler's family has more skateboards than Mike's family.

Luke's family has 1 fewer bicycle than Tyler's family.

a. Who is A? ______________ b. Who is B? ______________

c. Who is C? ______________ d. Who is D? ______________

Representations 2

School Trips

Developing methods for solving problems systematically

AIM

Students will use drawings, symbols, and tables to represent and solve real-world problems. They will also translate between different representations.

MATERIALS

- 1 copy of the blackline master (opposite) for each student

REFLECTION

Review the different ways to solve problems (diagrams, tables, and number sentences) and how each is related.

1 Say, *Some students are going to the zoo. Cars can each take 3 students and vans can each take 5 students. If there are 45 students going on the trip, how many ways can they travel to the zoo, filling every seat in the vehicles?* Call on volunteers to share their solutions. Ask, *How can we figure out this problem in a systematic way?* Discuss the following methods:

- Draw a diagram to solve the problem.

  Say, *If they used only vans, how many will they need?* (9) Use the letters V and C to represent the vehicles. Draw ***V V V V V V V V V*** on the board. Ask, *If they use 8 vans, how many cars will they need?* (2) *Will all the seats be occupied?* (No.) *Is this a solution?* (No. Not all the seats are occupied.) *If they use 7 vans, how many cars will they need?* (4) *Will all the seats be occupied?* (No.) *If they use 6 vans, how many cars will they need?* (5) *Will all the seats be occupied?* (Yes.) Continue until all possible solutions are shown.

- Use addition to solve the problem.

  Write ***5 + 5 + 5 + 5 + 5 + 5 + 5 + 5 + 5*** and ***5 + 5 + 5 + 5 + 5 + 5 + 3 + 3 + 3 + 3 + 3*** on the board. Continue until all possible solutions are shown.

- Draw a table, as shown below, to solve the problem.

Number of Vans	Number of Cars
9	0
6	3
3	6
0	15

Discuss how each representation is related, and how they are all systematic ways of finding all the possible solutions.

2 Ask the students to complete the blackline master. Call on volunteers to share their answers. Check for systematic answers, and discuss the pattern in the table.

School Trips

Name ______________________

The local park has 3-seater canoes and 2-seater canoes. The park will only rent out canoes if they are full. How many different ways can 36 students fit in the canoes?

1. Draw a diagram to show the solutions.

2. Complete this table.

3-seater canoes	2-seater canoes	Number of students
12	0	36
		36
		36
		36
		36
		36
		36

3. a. If you want 8 more 2-seater canoes than 3-seater canoes, how many of each will you rent?

2-seater canoes = ______

3-seater canoes = ______

b. If you want to have the fewest number of canoes, how many of each will you rent?

2-seater canoes = ______

3-seater canoes = ______

For the Class

Developing methods for solving problems systematically

AIM

Students will use drawings, symbols, and tables to represent and solve real-world problems. They will also translate between different representations.

MATERIALS

- 1 copy of the blackline master (opposite) for each student

REFLECTION

Review the different ways to solve real-world problems (diagrams, tables, and number sentences) and how each is related.

1 Say, *Pencils cost 50 cents each and erasers cost 20 cents each. How many pencils and erasers can we buy for $5, without change? Is there more than one answer? How can we figure out the possible answers systematically?* Discuss the following methods:

- Draw a diagram to solve the problem.

  Ask, *If we buy pencils only, how many can we buy?* Draw 10 pencils on the board. Cross out 1 pencil and ask, *If we buy 9 pencils, how many erasers can we buy? Will we get change?* (Yes, so it is not a solution.) Cross out 1 more pencil and ask, *How many erasers can we buy now?* (5) *Will we get change?* (No.) Draw 5 erasers. Continue until all possible solutions are shown.

- Use addition to solve the problem.

  Write ***50 + 50 + 50 + 50 + 50 + 50 + 50 + 50 + 50 + 50*** and ***50 + 50 + 50 + 50 + 50 + 50 + 50 + 50 + 20 + 20 + 20 + 20 + 20*** on the board. Continue, swapping pencils for erasers until all possible solutions are shown.

- Draw a table, as shown below, to solve the problem.

Number of pencils	Number of erasers
10	0
8	5
6	10
4	15
2	20
0	25

  Discuss how each representation is related, and how they are all systematic ways of finding all the solutions.

2 Read Question 1 on the blackline master with the students. Discuss which method would be best for showing all the possible solutions. Work with the class to construct a table. It could have 3 columns: Notebooks, Pencils, and Total cost. Allow time for the students to complete the table and answer Question 2.

For the Class

Name ______________________________

1. Imagine you have $30. Use the space below to figure out how many of each item you can buy without any money left over. There is more than one way.

2. a. If you want twice as many pencils as notebooks, how many of each will you buy?

   Notebooks = _____ Pencils = _____

   b. If you want the greatest number of items, how many of each will you buy?

   Notebooks = _____ Pencils = _____

Older Again

Plotting points on a grid

AIM

Students will enter data into a table and on a grid. They will then use the table and the grid to describe real-world problems.

MATERIALS

- 1 copy of the blackline master (opposite) for each student
- Overhead projector
- 1 copy of the blackline master (opposite) on an overhead transparency

REFLECTION

Discuss the grid in Question 2 on the blackline master. Ask, *If we know the children's ages now, how do we figure out their ages in 5 years?* (Add 5.) *If we know how old they will be in 5 years, how can we figure out their ages now?* (Subtract 5.) *What do you notice about the points on the graph?* (They are in a straight line.) Invite a volunteer to choose any new point on the same line, and ask, *If Gemma is plotted here, what do we know about her?*

1 Read Question 1 on the blackline master with the class. Copy the 1st table onto the board. Ask, *How old is Rashid now?* (12) *How old will he be in 5 years?* (17) Record the data in the table. Ask the students to complete Question 1. Call on volunteers to share their answers.

2 Show the overhead transparency. Refer to the grid. Ask, *What do the numbers on the horizontal axis show? What do the numbers on the vertical axis show? How old is Yoshi now? How do you know? How old will he be in 5 years? How do you know?* Call on volunteers to share their thinking. Ask, *How old is Rose now? How old will she be in 5 years?* Ask, *If Alex is 3 years old now, how old will he be in 5 years?* Invite a volunteer to plot a point showing Alex on the grid. Repeat for another child who is 10 years old now.

3 Have the students complete the blackline master. Ask, *What do you notice about the grid?* (The points are on a straight line.)

Older Again

Name ______________________

1. Complete the tables to show how old the children will be in 5 years.

	Age now	Age in 5 years
Rashid	12	
Reno	2	

	Age now	Age in 5 years
Lana	7	
Chan	5	

2. Plot their ages on the grid. Join all the points with a line.

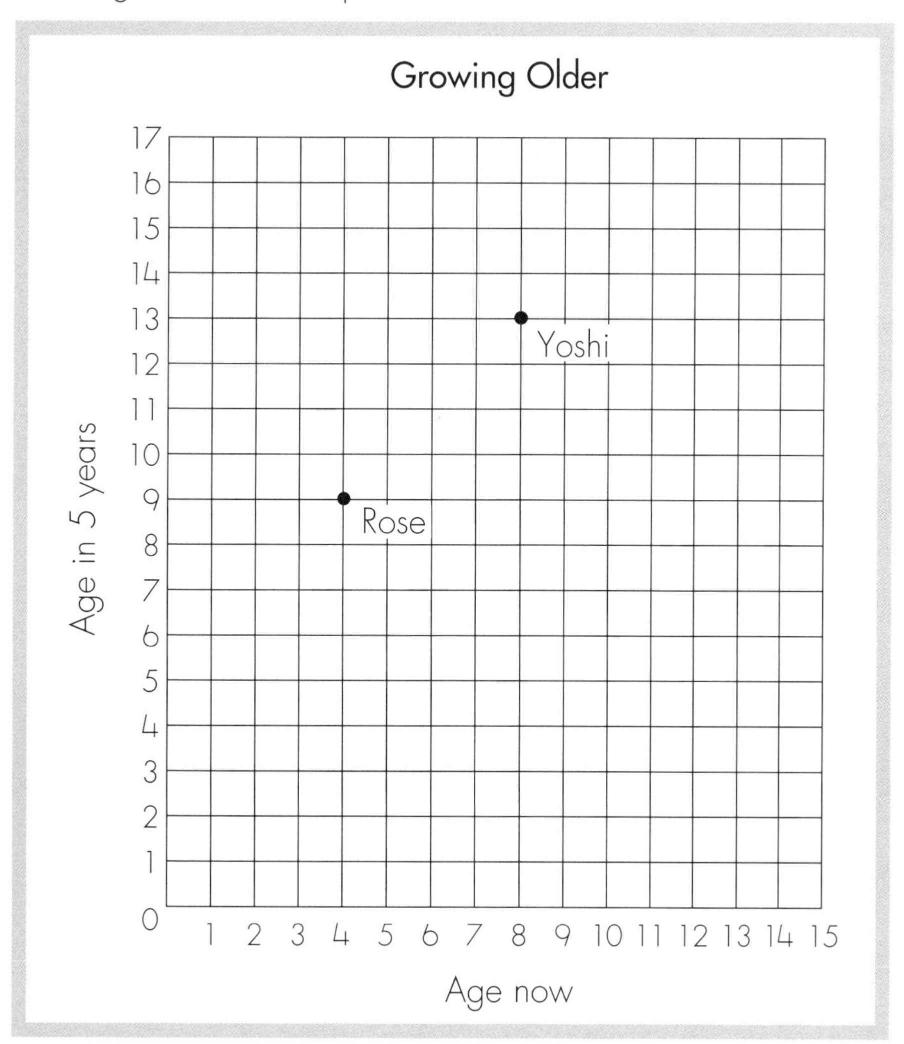

3. If Salma is going to be 16 years old in 5 years, how old is she now? ______
Plot a point showing Salma on the grid.

4. If Roberto is 9 years old now, how old will he be in 5 years? ______
Plot a point showing Roberto on the grid.

Representations 5

Toy Sale

Plotting points on a grid

AIM

Students will enter data into a table and on a grid. They will then use the table and the grid to describe real-world problems.

MATERIALS

- 1 copy of the blackline master (opposite) for each student
- Overhead projector
- 1 copy of the blackline master (opposite) on an overhead transparency

REFLECTION

Refer to the grid on the blackline master and ask, *If we know the original price, how do we figure out the sale price?* (Subtract 3.) *If we know the sale price, how do we figure out the original price?* (Add 3.) *What do you notice about the points on the grid?* (They are in a straight line.) *If a toy costs $5.50 before the sale, where will this point be on the grid? If a toy has a sale price of $21.50, where will this point be on the grid?*

1 Read Question 1 on the blackline master with the class. Ask, *How many items are on sale? How much less do they cost?* Ask the students to complete the table and share their answers.

2 Show the overhead transparency. Refer to the grid. Ask, *What do the numbers on the horizontal axis show? What do the numbers on the vertical axis show? What is the original price of beach balls?* ($14) *What is their sale price?* ($11) *What is the original price of kites?* ($6) *What is their sale price?* ($3) Refer to the table in Question 1 and ask, *What is the original price of the blocks? Where will they be on the horizontal axis? What is their sale price? Where will they be on the vertical axis? Where do we plot the blocks on the grid?* Together, plot a point showing the set of blocks. Ask, *What is the original price of model cars? What is their sale price?* Together, plot a point showing the model cars.

3 Ask the students to complete the blackline master. Call on volunteers to share their responses.

Toy Sale

Name ______________________

1. Toys are on sale. They now cost $3 less. Complete this table.

	Blocks	Dolls	Puzzles	Footballs	Model cars
Original price	$25	$8	$9	$15	$5
Sale price					

2. Plot a point showing each toy on the grid. Join all the points with a line.

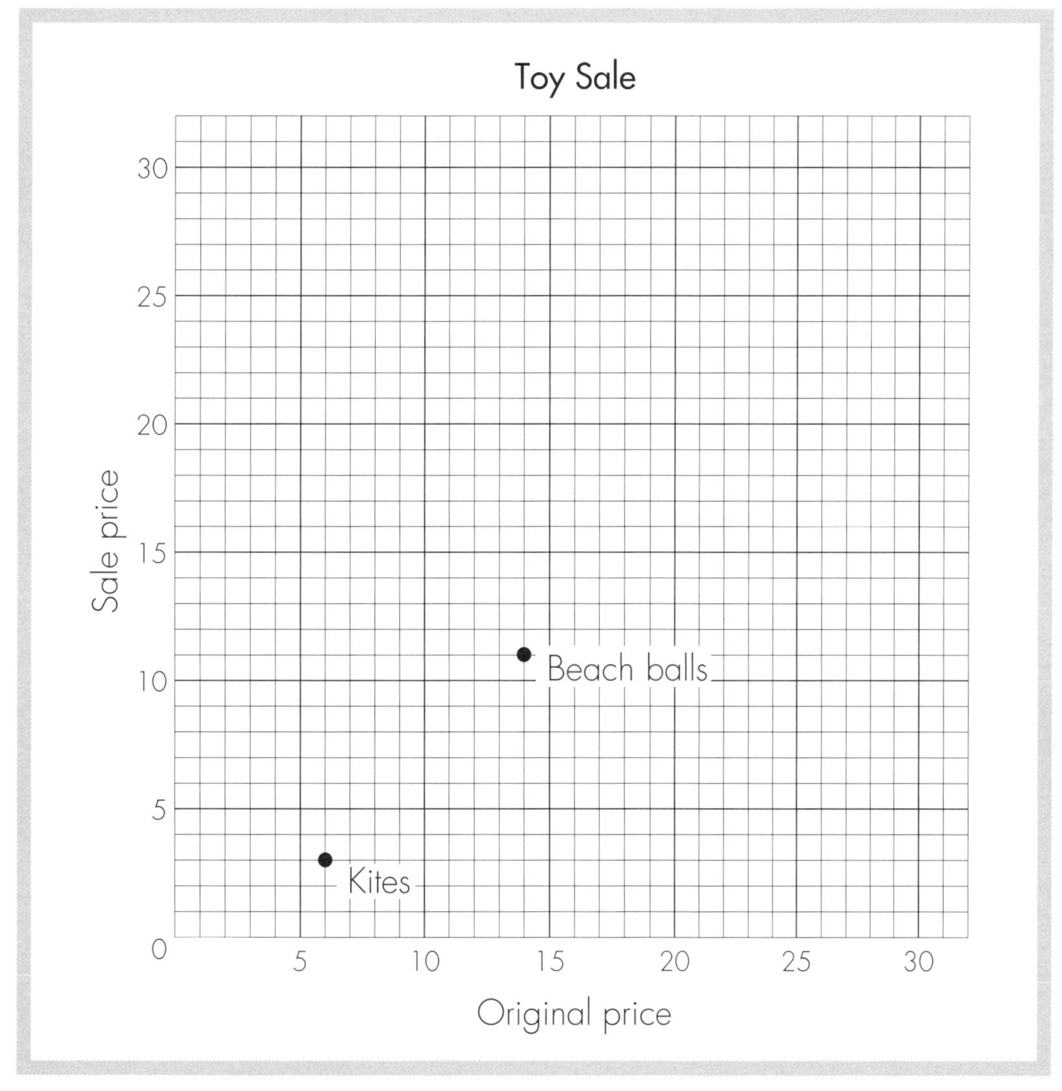

3. Look at the grid and complete the sentences.

   a. The original price of beach balls is ________.

   b. The sale price of kites is ________.

Representations 6

ANSWERS

Equivalence and Equations 1 — Page 7

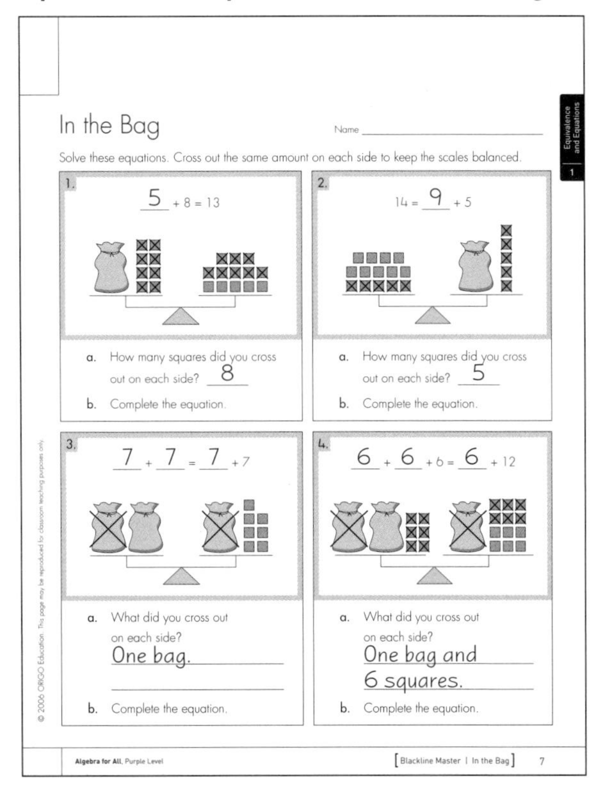

In the Bag

Name

Solve these equations. Cross out the same amount on each side to keep the scales balanced.

1. 5 + 8 = 13
   a. How many squares did you cross out on each side? 8
   b. Complete the equation.
2. 14 = 9 + 5
   a. How many squares did you cross out on each side? 5
   b. Complete the equation.
3. 7 + 7 = 7 + 7
   a. What did you cross out on each side? One bag.
   b. Complete the equation.
4. 6 + 6 + 6 = 6 + 12
   a. What did you cross out on each side? One bag and 6 squares.
   b. Complete the equation.

Equivalence and Equations 1

© 2006 ORIGO Education. This page may be reproduced for classroom teaching purposes only.

Algebra for All, Purple Level [Blackline Master | In the Bag] 7

Equivalence and Equations 2 — Page 9

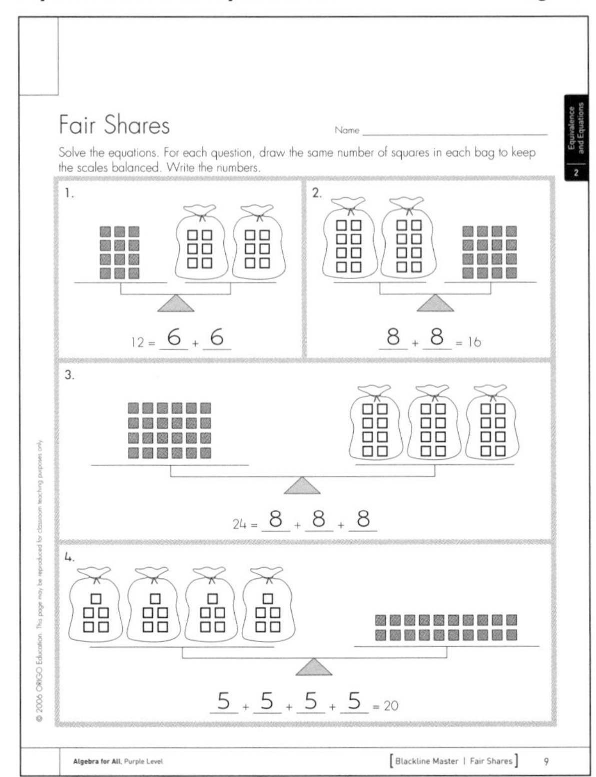

Fair Shares

Name

Solve the equations. For each question, draw the same number of squares in each bag to keep the scales balanced. Write the numbers.

1. 12 = 6 + 6
2. 8 + 8 = 16
3. 24 = 8 + 8 + 8
4. 5 + 5 + 5 + 5 = 20

Equivalence and Equations 2

© 2006 ORIGO Education. This page may be reproduced for classroom teaching purposes only.

Algebra for All, Purple Level [Blackline Master | Fair Shares] 9

Equivalence and Equations 3 — Page 11

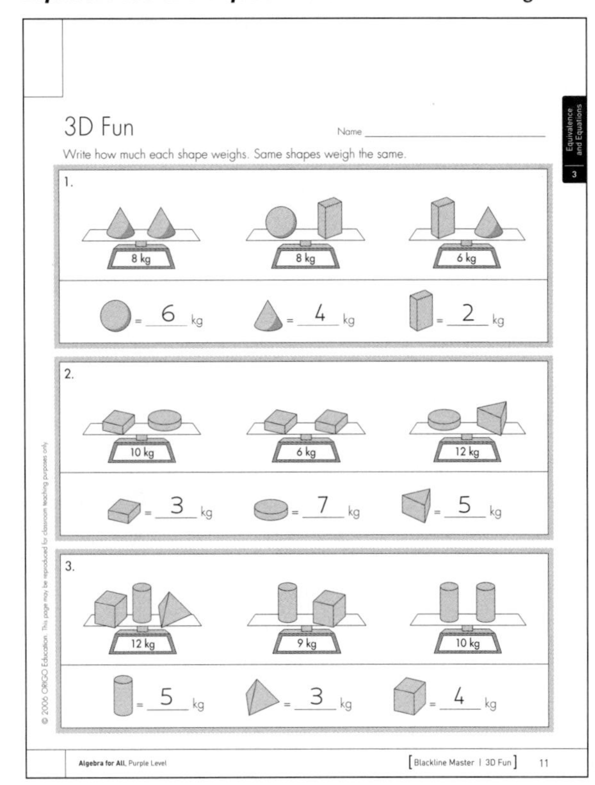

3D Fun

Name

Write how much each shape weighs. Same shapes weigh the same.

1. 8 kg, 8 kg, 6 kg — sphere = 6 kg, cone = 4 kg, prism = 2 kg
2. 10 kg, 6 kg, 12 kg — block = 3 kg, disc = 7 kg, wedge = 5 kg
3. 12 kg, 9 kg, 10 kg — cylinder = 5 kg, pyramid = 3 kg, cube = 4 kg

Equivalence and Equations 3

© 2006 ORIGO Education. This page may be reproduced for classroom teaching purposes only.

Algebra for All, Purple Level [Blackline Master | 3D Fun] 11

Equivalence and Equations 4 — Page 13

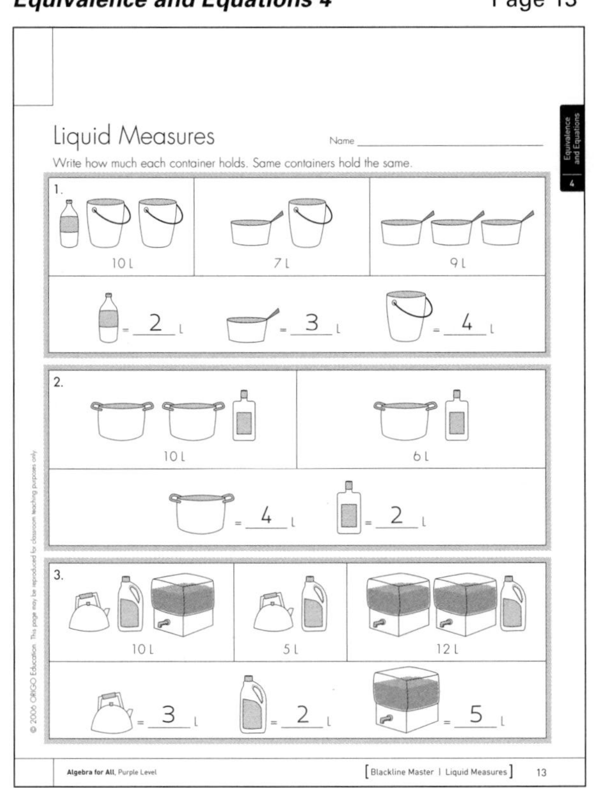

Liquid Measures

Name

Write how much each container holds. Same containers hold the same.

1. 10 L, 7 L, 9 L — bottle = 2 L, pan = 3 L, bucket = 4 L
2. 10 L, 6 L — pot = 4 L, bottle = 2 L
3. 10 L, 5 L, 12 L — kettle = 3 L, bottle = 2 L, tank = 5 L

Equivalence and Equations 4

© 2006 ORIGO Education. This page may be reproduced for classroom teaching purposes only.

Algebra for All, Purple Level [Blackline Master | Liquid Measures] 13

Equivalence and Equations 5 Page 15

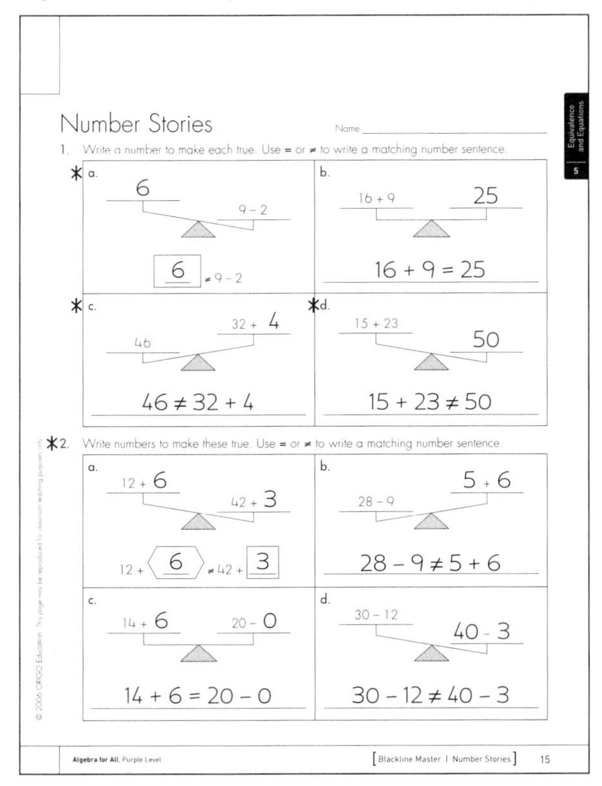
Number Stories Name

1. Write a number to make each true. Use = or ≠ to write a matching number sentence.

*a. 6 | 9 − 2 — 6 ≠ 9 − 2

b. 16 + 9 | 25 — 16 + 9 = 25

*c. 46 | 32 + 4 — 46 ≠ 32 + 4

*d. 15 + 23 | 50 — 15 + 23 ≠ 50

*2. Write numbers to make these true. Use = or ≠ to write a matching number sentence.

a. 12 + 6 | 42 + 3 — 12 + 6 ≠ 42 + 3

b. 28 − 9 | 5 + 6 — 28 − 9 ≠ 5 + 6

c. 14 + 6 | 20 − 0 — 14 + 6 = 20 − 0

d. 30 − 12 | 40 − 3 — 30 − 12 ≠ 40 − 3

Algebra for All, Purple Level [Blackline Master | Number Stories] 15

Equivalence and Equations 6 Page 17

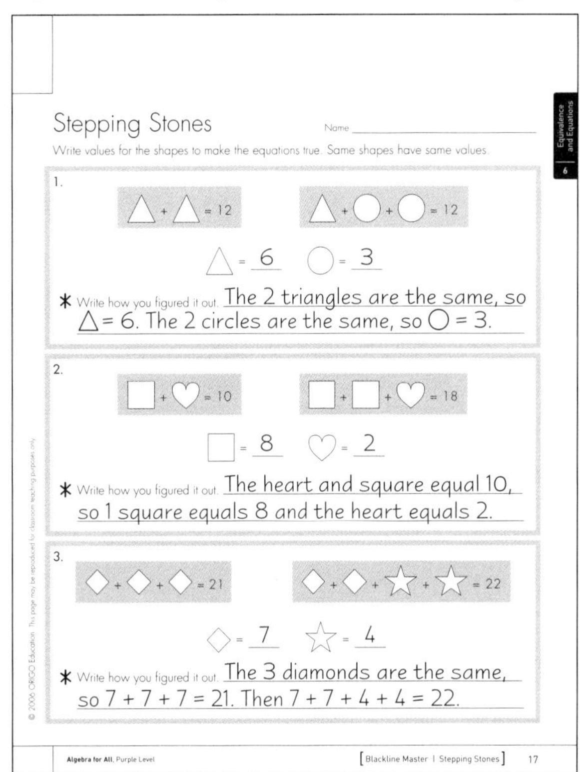
Stepping Stones Name

Write values for the shapes to make the equations true. Same shapes have same values.

1. △ + △ = 12 △ + ○ + ○ = 12

△ = 6 ○ = 3

* Write how you figured it out. The 2 triangles are the same, so △ = 6. The 2 circles are the same, so ○ = 3.

2. □ + ♡ = 10 □ + □ + ♡ = 18

□ = 8 ♡ = 2

* Write how you figured it out. The heart and square equal 10, so 1 square equals 8 and the heart equals 2.

3. ◇ + ◇ + ◇ = 21 ◇ + ◇ + ☆ + ☆ = 22

◇ = 7 ☆ = 4

* Write how you figured it out. The 3 diamonds are the same, so 7 + 7 + 7 = 21. Then 7 + 7 + 4 + 4 = 22.

Algebra for All, Purple Level [Blackline Master | Stepping Stones] 17

Equivalence and Equations 7 Page 19

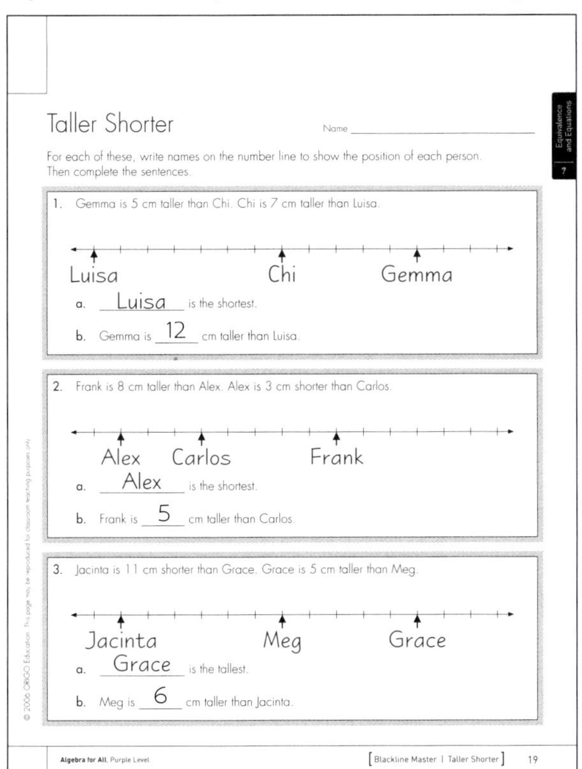
Taller Shorter Name

For each of these, write names on the number line to show the position of each person. Then complete the sentences.

1. Gemma is 5 cm taller than Chi. Chi is 7 cm taller than Luisa.

Luisa Chi Gemma

a. Luisa is the shortest.

b. Gemma is 12 cm taller than Luisa.

2. Frank is 8 cm taller than Alex. Alex is 3 cm shorter than Carlos.

Alex Carlos Frank

a. Alex is the shortest.

b. Frank is 5 cm taller than Carlos.

3. Jacinta is 11 cm shorter than Grace. Grace is 5 cm taller than Meg.

Jacinta Meg Grace

a. Grace is the tallest.

b. Meg is 6 cm taller than Jacinta.

Algebra for All, Purple Level [Blackline Master | Taller Shorter] 19

Equivalence and Equations 8 Page 21

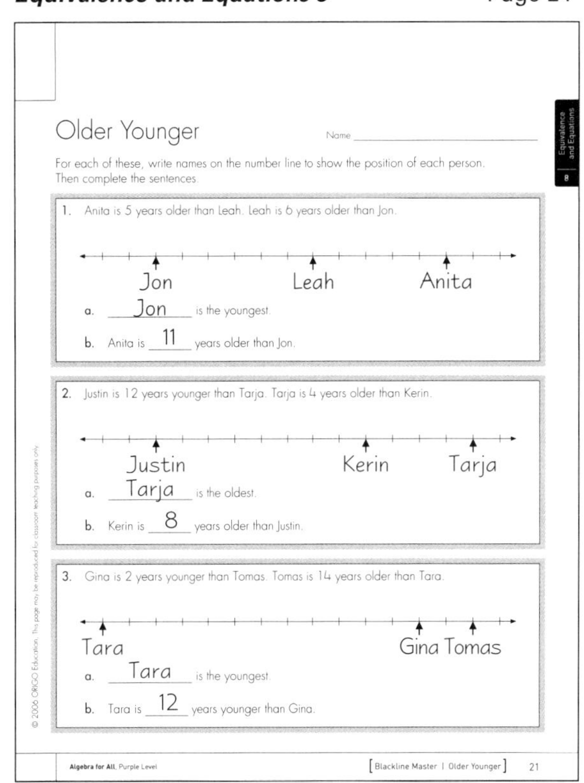
Older Younger Name

For each of these, write names on the number line to show the position of each person. Then complete the sentences.

1. Anita is 5 years older than Leah. Leah is 6 years older than Jon.

Jon Leah Anita

a. Jon is the youngest.

b. Anita is 11 years older than Jon.

2. Justin is 12 years younger than Tarja. Tarja is 4 years older than Kerin.

Justin Kerin Tarja

a. Tarja is the oldest.

b. Kerin is 8 years older than Justin.

3. Gina is 2 years younger than Tomas. Tomas is 14 years older than Tara.

Tara Gina Tomas

a. Tara is the youngest.

b. Tara is 12 years younger than Gina.

Algebra for All, Purple Level [Blackline Master | Older Younger] 21

* Answers will vary. This is one example.

ANSWERS

Patterns and Functions 1 — Page 23

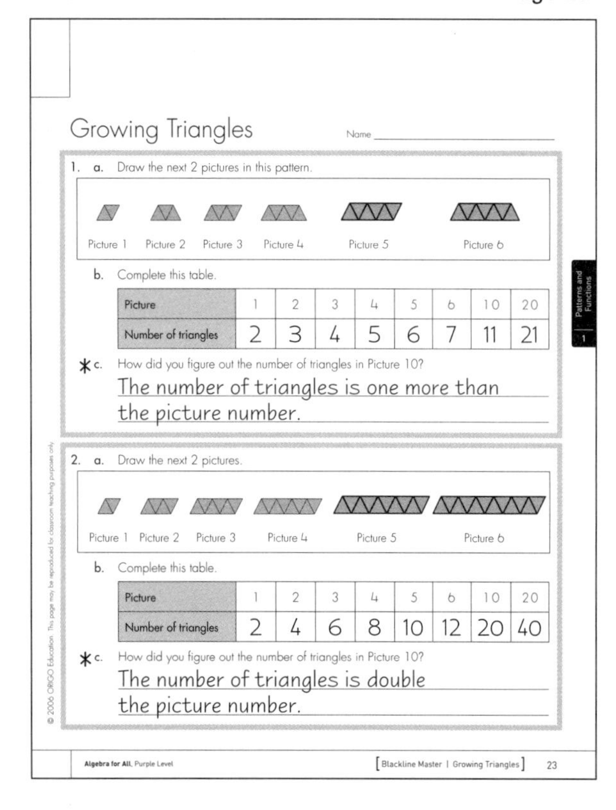

Growing Triangles

Name

1. a. Draw the next 2 pictures in this pattern.

Picture 1 Picture 2 Picture 3 Picture 4 Picture 5 Picture 6

b. Complete this table.

Picture	1	2	3	4	5	6	10	20
Number of triangles	2	3	4	5	6	7	11	21

*c. How did you figure out the number of triangles in Picture 10?
The number of triangles is one more than the picture number.

2. a. Draw the next 2 pictures.

Picture 1 Picture 2 Picture 3 Picture 4 Picture 5 Picture 6

b. Complete this table.

Picture	1	2	3	4	5	6	10	20
Number of triangles	2	4	6	8	10	12	20	40

*c. How did you figure out the number of triangles in Picture 10?
The number of triangles is double the picture number.

Patterns and Functions 1

© 2006 ORIGO Education. This page may be reproduced for classroom teaching purposes only.

Algebra for All, Purple Level [Blackline Master | Growing Triangles] 23

Patterns and Functions 2 — Page 25

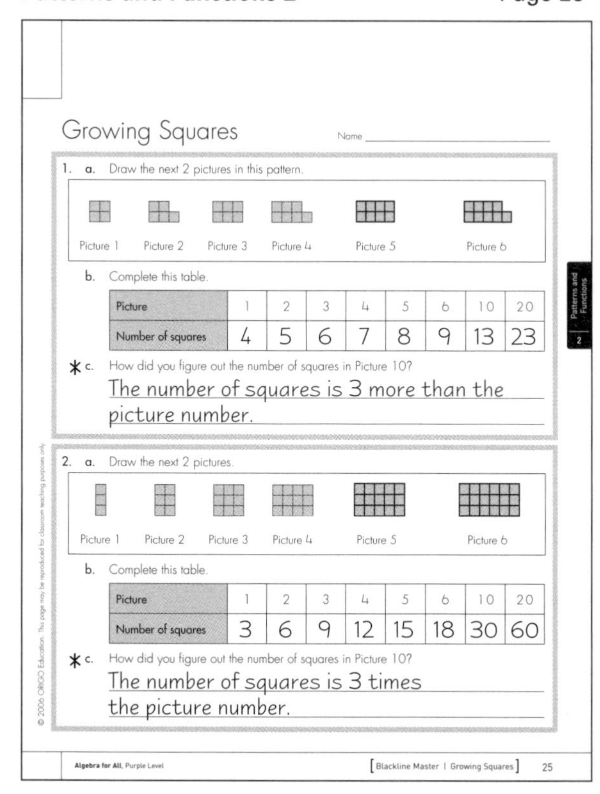

Growing Squares

Name

1. a. Draw the next 2 pictures in this pattern.

Picture 1 Picture 2 Picture 3 Picture 4 Picture 5 Picture 6

b. Complete this table.

Picture	1	2	3	4	5	6	10	20
Number of squares	4	5	6	7	8	9	13	23

*c. How did you figure out the number of squares in Picture 10?
The number of squares is 3 more than the picture number.

2. a. Draw the next 2 pictures.

Picture 1 Picture 2 Picture 3 Picture 4 Picture 5 Picture 6

b. Complete this table.

Picture	1	2	3	4	5	6	10	20
Number of squares	3	6	9	12	15	18	30	60

*c. How did you figure out the number of squares in Picture 10?
The number of squares is 3 times the picture number.

Patterns and Functions 2

© 2006 ORIGO Education. This page may be reproduced for classroom teaching purposes only.

Algebra for All, Purple Level [Blackline Master | Growing Squares] 25

Patterns and Functions 3 — Page 27

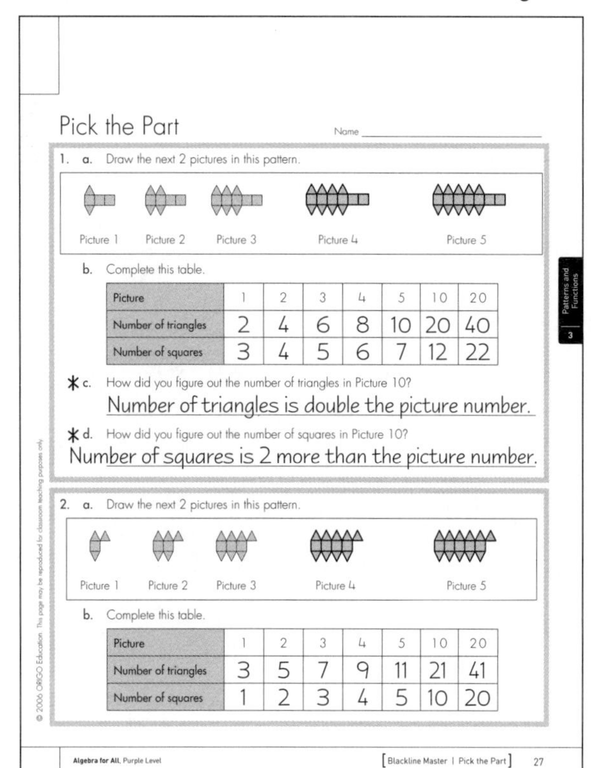

Pick the Part

Name

1. a. Draw the next 2 pictures in this pattern.

Picture 1 Picture 2 Picture 3 Picture 4 Picture 5

b. Complete this table.

Picture	1	2	3	4	5	10	20
Number of triangles	2	4	6	8	10	20	40
Number of squares	3	4	5	6	7	12	22

*c. How did you figure out the number of triangles in Picture 10?
Number of triangles is double the picture number.

*d. How did you figure out the number of squares in Picture 10?
Number of squares is 2 more than the picture number.

2. a. Draw the next 2 pictures in this pattern.

Picture 1 Picture 2 Picture 3 Picture 4 Picture 5

b. Complete this table.

Picture	1	2	3	4	5	10	20
Number of triangles	3	5	7	9	11	21	41
Number of squares	1	2	3	4	5	10	20

Patterns and Functions 3

© 2006 ORIGO Education. This page may be reproduced for classroom teaching purposes only.

Algebra for All, Purple Level [Blackline Master | Pick the Part] 27

Patterns and Functions 4 — Page 29

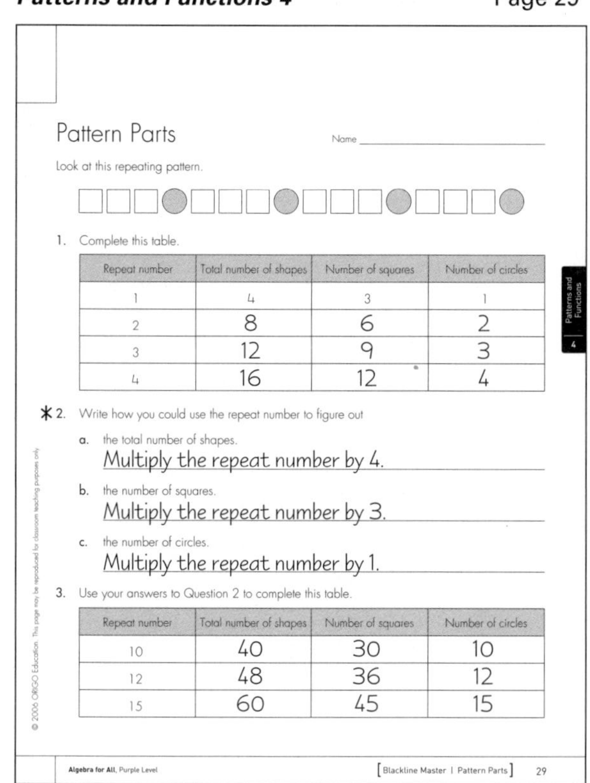

Pattern Parts

Name

Look at this repeating pattern.

1. Complete this table.

Repeat number	Total number of shapes	Number of squares	Number of circles
1	4	3	1
2	8	6	2
3	12	9	3
4	16	12	4

*2. Write how you could use the repeat number to figure out

a. the total number of shapes.
Multiply the repeat number by 4.

b. the number of squares.
Multiply the repeat number by 3.

c. the number of circles.
Multiply the repeat number by 1.

3. Use your answers to Question 2 to complete this table.

Repeat number	Total number of shapes	Number of squares	Number of circles
10	40	30	10
12	48	36	12
15	60	45	15

Patterns and Functions 4

© 2006 ORIGO Education. This page may be reproduced for classroom teaching purposes only.

Algebra for All, Purple Level [Blackline Master | Pattern Parts] 29

* Answers will vary. This is one example.

Patterns and Functions 5

Page 31

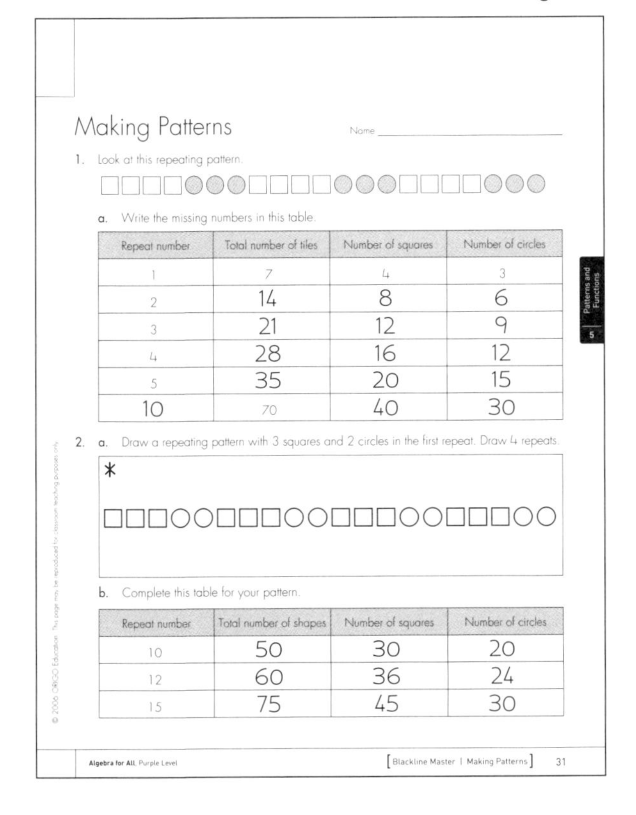

Making Patterns

Name

1. Look at this repeating pattern.

□□□□○○○□□□□○○○□□□□○○○

a. Write the missing numbers in this table.

Repeat number	Total number of tiles	Number of squares	Number of circles
1	7	4	3
2	14	8	6
3	21	12	9
4	28	16	12
5	35	20	15
10	70	40	30

2. a. Draw a repeating pattern with 3 squares and 2 circles in the first repeat. Draw 4 repeats.

*

□□□○○□□□○○□□□○○□□□○○

b. Complete this table for your pattern.

Repeat number	Total number of shapes	Number of squares	Number of circles
10	50	30	20
12	60	36	24
15	75	45	30

Patterns and Functions 5

Patterns and Functions 6

Page 33

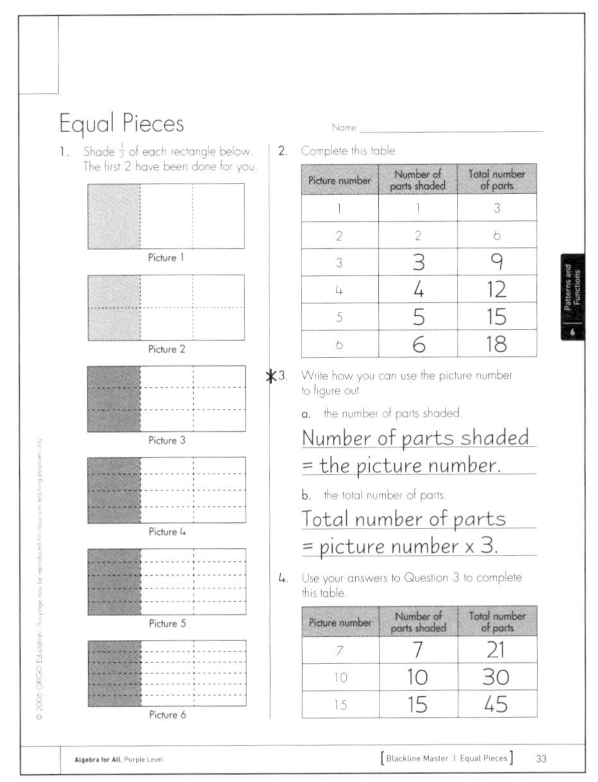

Equal Pieces

Name

1. Shade $\frac{1}{3}$ of each rectangle below. The first 2 have been done for you.

Picture 1

Picture 2

Picture 3

Picture 4

Picture 5

Picture 6

2. Complete this table.

Picture number	Number of parts shaded	Total number of parts
1	1	3
2	2	6
3	3	9
4	4	12
5	5	15
6	6	18

*3. Write how you can use the picture number to figure out

a. the number of parts shaded.

Number of parts shaded = the picture number.

b. the total number of parts.

Total number of parts = picture number x 3.

4. Use your answers to Question 3 to complete this table.

Picture number	Number of parts shaded	Total number of parts
7	7	21
10	10	30
15	15	45

Patterns and Functions 6

Patterns and Functions 7

Page 35

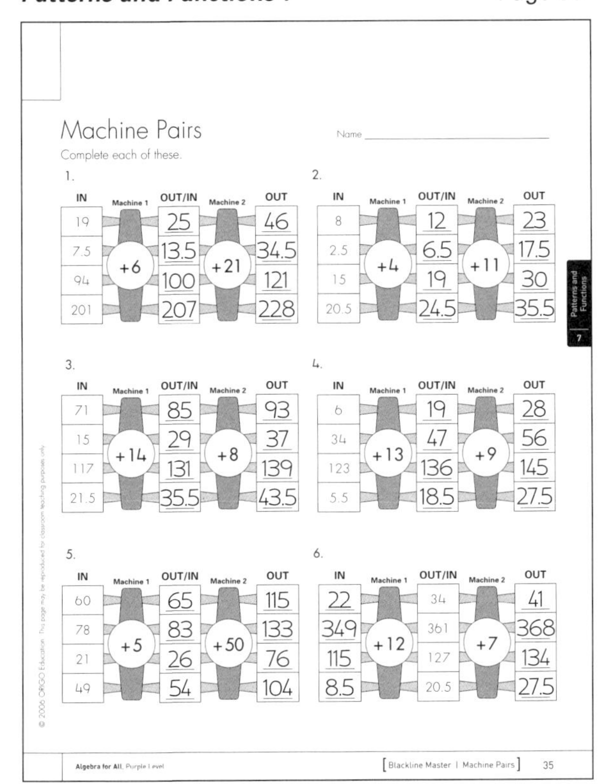

Machine Pairs

Name

Complete each of these.

1. Machine 1: +6; Machine 2: +21

IN	OUT/IN	OUT
19	25	46
7.5	13.5	34.5
94	100	121
201	207	228

2. Machine 1: +4; Machine 2: +11

IN	OUT/IN	OUT
8	12	23
2.5	6.5	17.5
15	19	30
20.5	24.5	35.5

3. Machine 1: +14; Machine 2: +8

IN	OUT/IN	OUT
71	85	93
15	29	37
117	131	139
21.5	35.5	43.5

4. Machine 1: +13; Machine 2: +9

IN	OUT/IN	OUT
6	19	28
34	47	56
123	136	145
5.5	18.5	27.5

5. Machine 1: +5; Machine 2: +50

IN	OUT/IN	OUT
60	65	115
78	83	133
21	26	76
49	54	104

6. Machine 1: +12; Machine 2: +7

IN	OUT/IN	OUT
22	34	41
349	361	368
115	127	134
8.5	20.5	27.5

Patterns and Functions 7

Patterns and Functions 8

Page 37

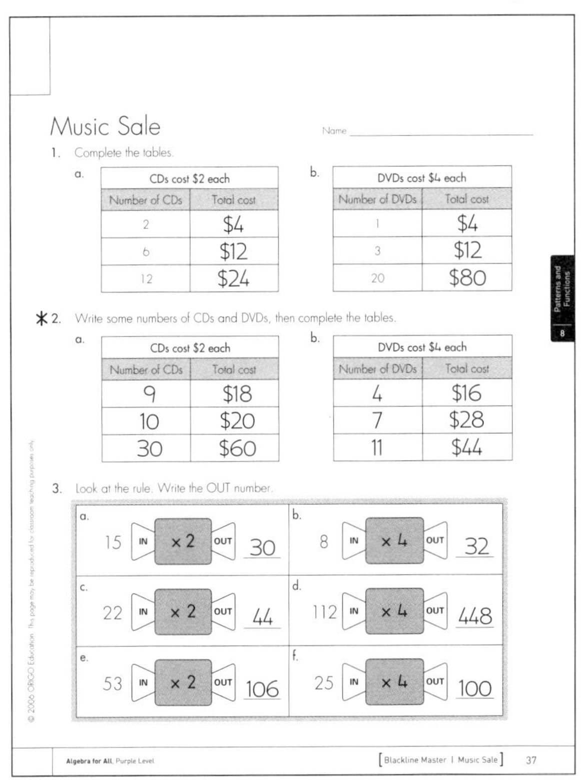

Music Sale

Name

1. Complete the tables.

a. CDs cost $2 each

Number of CDs	Total cost
2	$4
6	$12
12	$24

b. DVDs cost $4 each

Number of DVDs	Total cost
1	$4
3	$12
20	$80

*2. Write some numbers of CDs and DVDs, then complete the tables.

a. CDs cost $2 each

Number of CDs	Total cost
9	$18
10	$20
30	$60

b. DVDs cost $4 each

Number of DVDs	Total cost
4	$16
7	$28
11	$44

3. Look at the rule. Write the OUT number.

a. 15 IN x 2 OUT 30

b. 8 IN x 4 OUT 32

c. 22 IN x 2 OUT 44

d. 112 IN x 4 OUT 448

e. 53 IN x 2 OUT 106

f. 25 IN x 4 OUT 100

Patterns and Functions 8

* Answers will vary. This is one example.

ANSWERS

Patterns and Functions 9 — Page 39

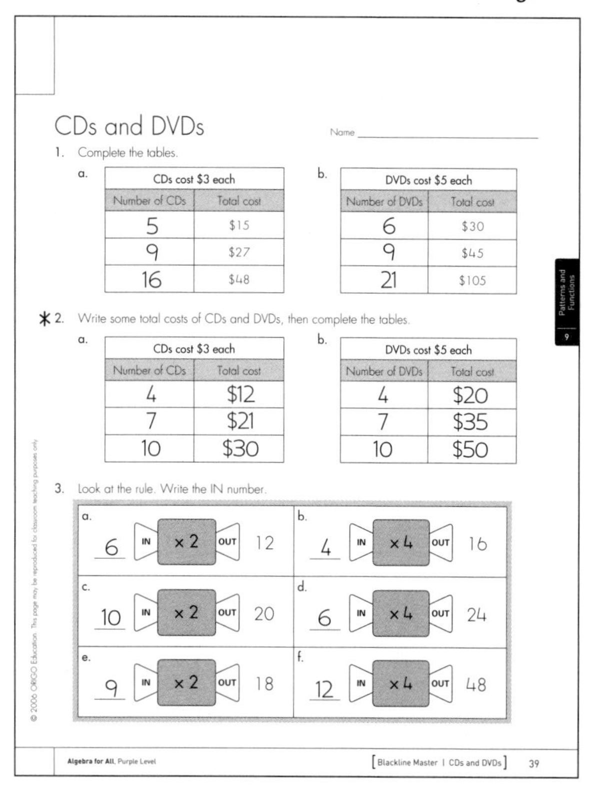

Patterns and Functions 10 — Page 41

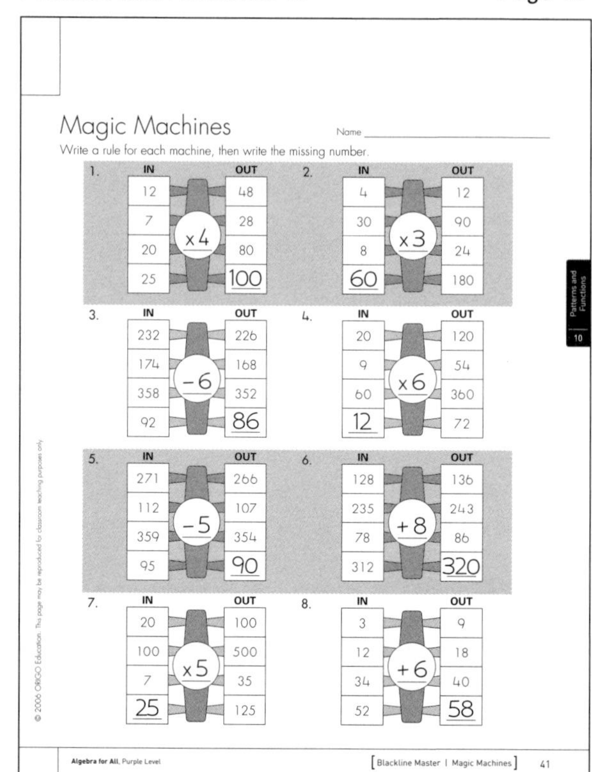

Patterns and Functions 11 — Page 43

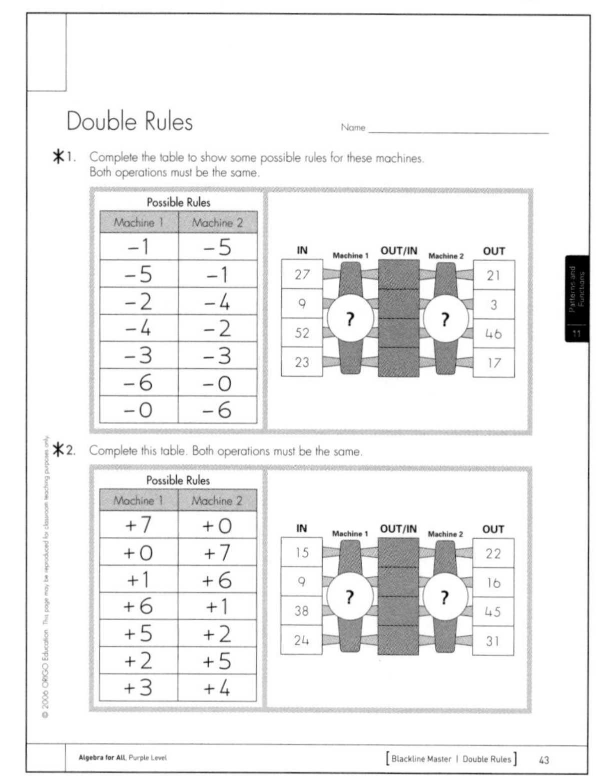

Properties 1 — Page 45

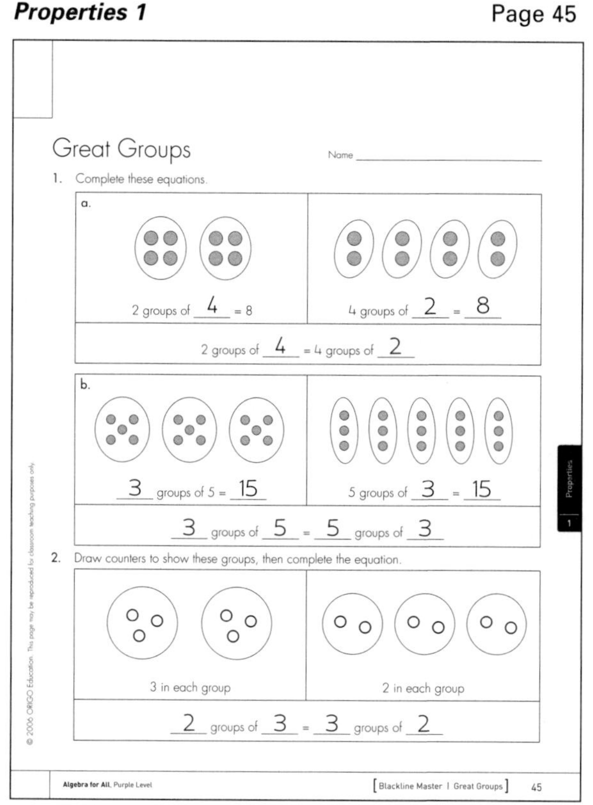

* Answers will vary. This is one example.

Properties 2 Page 47

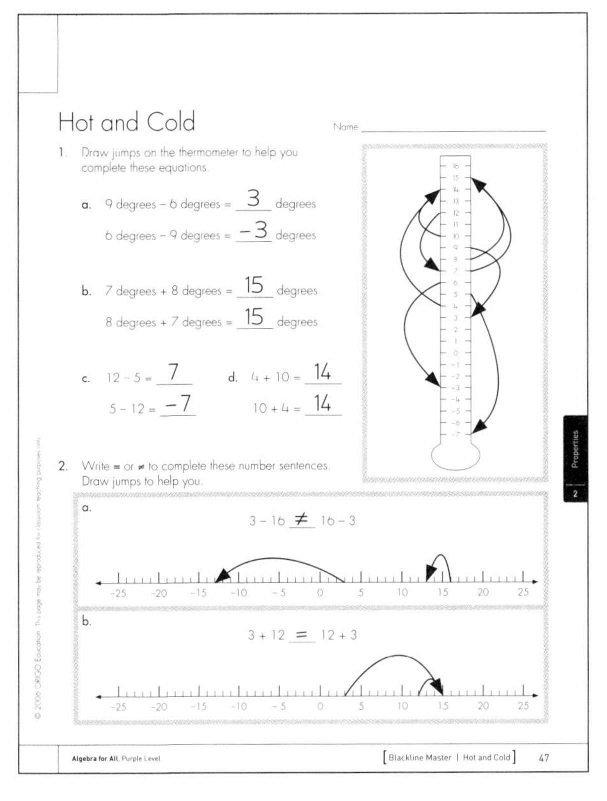

Hot and Cold

Name

1. Draw jumps on the thermometer to help you complete these equations.

a. 9 degrees − 6 degrees = 3 degrees

6 degrees − 9 degrees = −3 degrees

b. 7 degrees + 8 degrees = 15 degrees

8 degrees + 7 degrees = 15 degrees

c. 12 − 5 = 7

5 − 12 = −7

d. 4 + 10 = 14

10 + 4 = 14

2. Write = or ≠ to complete these number sentences. Draw jumps to help you.

a. 3 − 16 ≠ 16 − 3

b. 3 + 12 = 12 + 3

Algebra for All, Purple Level [Blackline Master | Hot and Cold] 47

Properties 3 Page 49

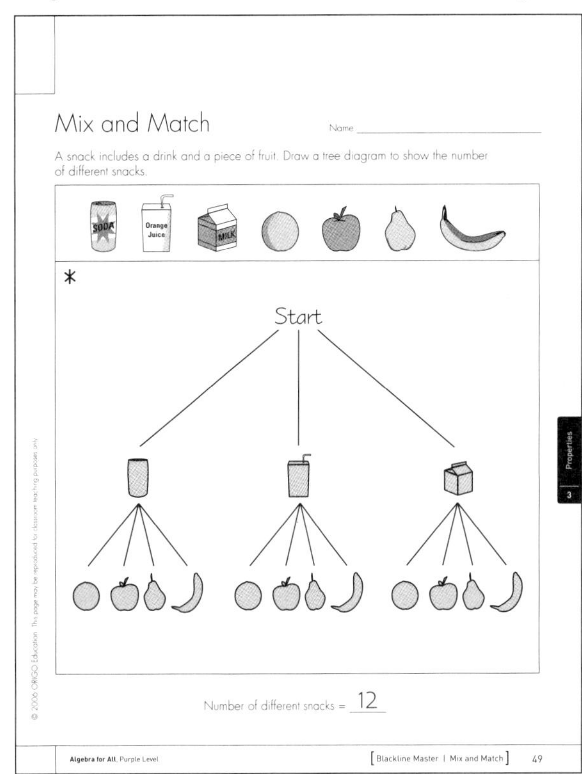

Mix and Match

Name

A snack includes a drink and a piece of fruit. Draw a tree diagram to show the number of different snacks.

Number of different snacks = 12

Algebra for All, Purple Level [Blackline Master | Mix and Match] 49

Properties 4 Page 51

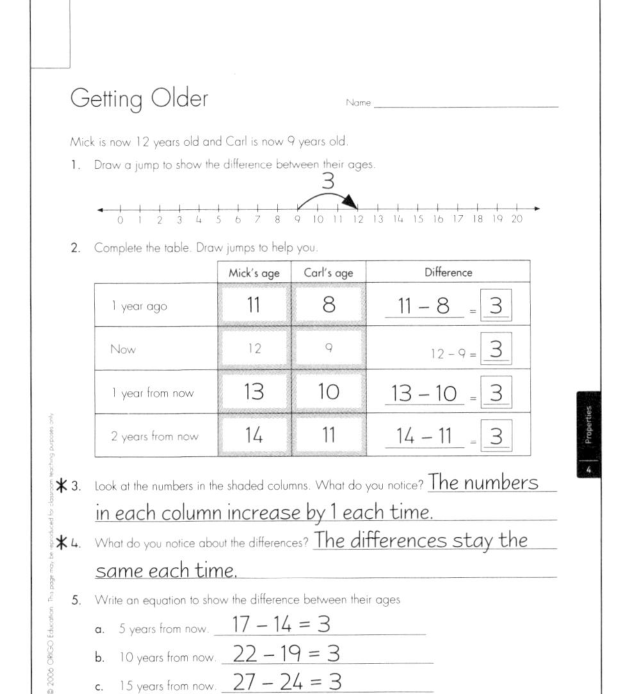

Getting Older

Name

Mick is now 12 years old and Carl is now 9 years old.

1. Draw a jump to show the difference between their ages.

2. Complete the table. Draw jumps to help you.

	Mick's age	Carl's age	Difference
1 year ago	11	8	11 − 8 = 3
Now	12	9	12 − 9 = 3
1 year from now	13	10	13 − 10 = 3
2 years from now	14	11	14 − 11 = 3

*3. Look at the numbers in the shaded columns. What do you notice? The numbers in each column increase by 1 each time.

*4. What do you notice about the differences? The differences stay the same each time.

5. Write an equation to show the difference between their ages.

a. 5 years from now 17 − 14 = 3

b. 10 years from now 22 − 19 = 3

c. 15 years from now 27 − 24 = 3

Algebra for All, Purple Level [Blackline Master | Getting Older] 51

Properties 5 Page 53

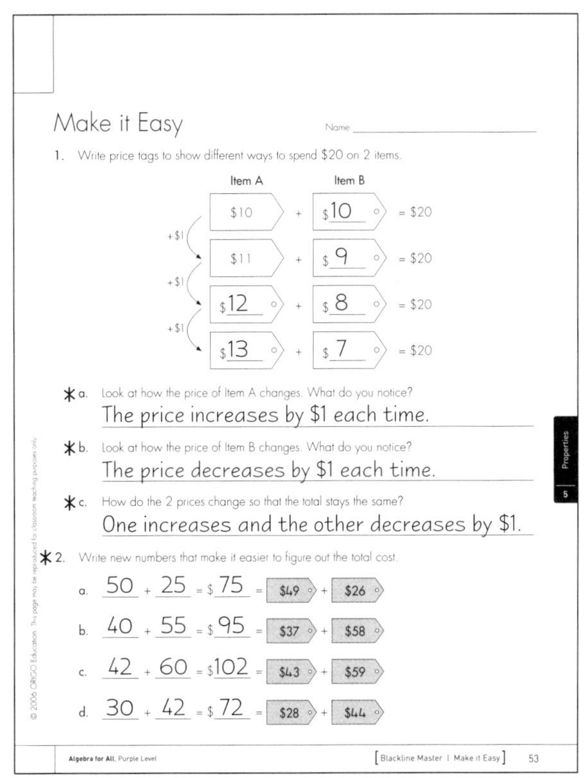

Make it Easy

Name

1. Write price tags to show different ways to spend $20 on 2 items.

Item A		Item B	
$10	+	$10	= $20
$11	+	$9	= $20
$12	+	$8	= $20
$13	+	$7	= $20

*a. Look at how the price of Item A changes. What do you notice? The price increases by $1 each time.

*b. Look at how the price of Item B changes. What do you notice? The price decreases by $1 each time.

*c. How do the 2 prices change so that the total stays the same? One increases and the other decreases by $1.

*2. Write new numbers that make it easier to figure out the total cost.

a. 50 + 25 = $75 = $49 + $26

b. 40 + 55 = $95 = $37 + $58

c. 42 + 60 = $102 = $43 + $59

d. 30 + 42 = $72 = $28 + $44

Algebra for All, Purple Level [Blackline Master | Make it Easy] 53

* Answers will vary. This is one example.

ANSWERS

Representations 1 — Page 55

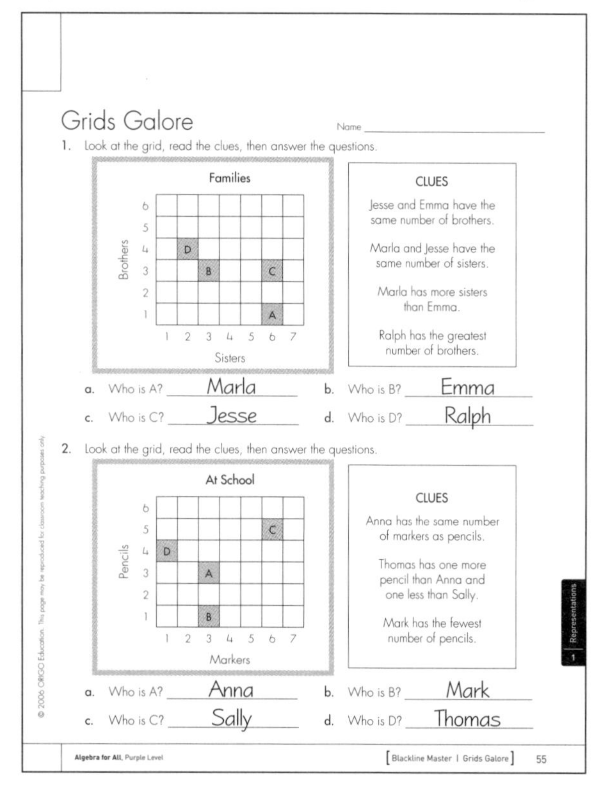

Grids Galore

Name ____________

1. Look at the grid, read the clues, then answer the questions.

CLUES

Jesse and Emma have the same number of brothers.

Marla and Jesse have the same number of sisters.

Marla has more sisters than Emma.

Ralph has the greatest number of brothers.

a. Who is A? Marla b. Who is B? Emma

c. Who is C? Jesse d. Who is D? Ralph

2. Look at the grid, read the clues, then answer the questions.

CLUES

Anna has the same number of markers as pencils.

Thomas has one more pencil than Anna and one less than Sally.

Mark has the fewest number of pencils.

a. Who is A? Anna b. Who is B? Mark

c. Who is C? Sally d. Who is D? Thomas

© 2006 ORIGO Education. This page may be reproduced for classroom teaching purposes only.

Algebra for All, Purple Level [Blackline Master | Grids Galore] 55

Representations 2 — Page 57

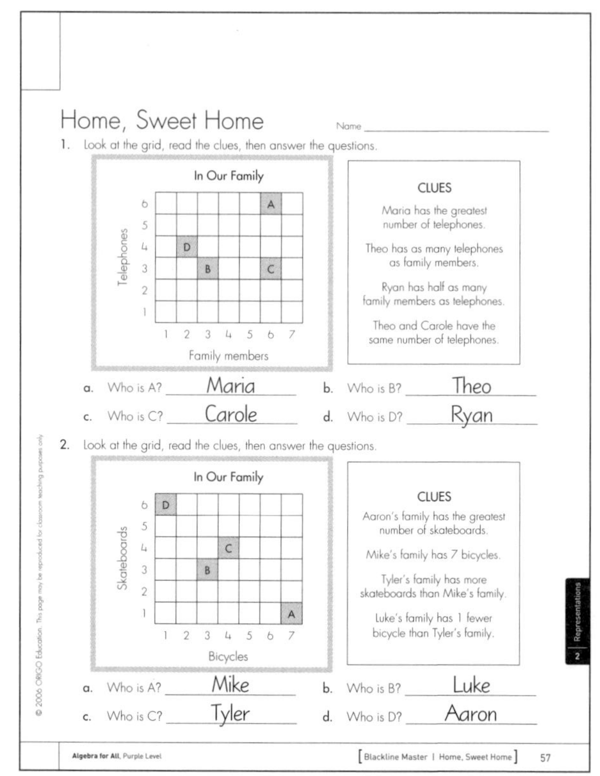

Home, Sweet Home

Name ____________

1. Look at the grid, read the clues, then answer the questions.

CLUES

Maria has the greatest number of telephones.

Theo has as many telephones as family members.

Ryan has half as many family members as telephones.

Theo and Carole have the same number of telephones.

a. Who is A? Maria b. Who is B? Theo

c. Who is C? Carole d. Who is D? Ryan

2. Look at the grid, read the clues, then answer the questions.

CLUES

Aaron's family has the greatest number of skateboards.

Mike's family has 7 bicycles.

Tyler's family has more skateboards than Mike's family.

Luke's family has 1 fewer bicycle than Tyler's family.

a. Who is A? Mike b. Who is B? Luke

c. Who is C? Tyler d. Who is D? Aaron

© 2006 ORIGO Education. This page may be reproduced for classroom teaching purposes only.

Algebra for All, Purple Level [Blackline Master | Home, Sweet Home] 57

Representations 3 — Page 59

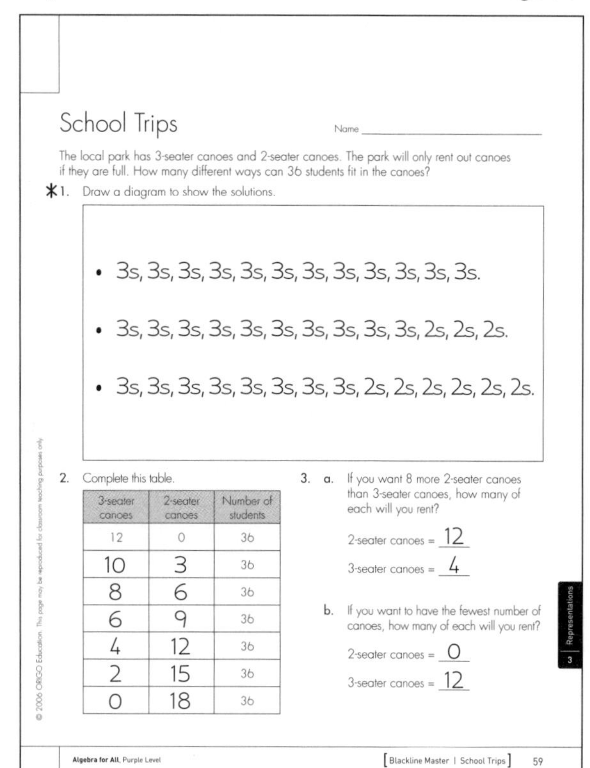

School Trips

Name ____________

The local park has 3-seater canoes and 2-seater canoes. The park will only rent out canoes if they are full. How many different ways can 36 students fit in the canoes?

*1. Draw a diagram to show the solutions.

- 3s, 3s, 3s, 3s, 3s, 3s, 3s, 3s, 3s, 3s, 3s, 3s.
- 3s, 3s, 3s, 3s, 3s, 3s, 3s, 3s, 3s, 3s, 2s, 2s, 2s.
- 3s, 3s, 3s, 3s, 3s, 3s, 3s, 3s, 2s, 2s, 2s, 2s, 2s, 2s.

2. Complete this table.

3-seater canoes	2-seater canoes	Number of students
12	0	36
10	3	36
8	6	36
6	9	36
4	12	36
2	15	36
0	18	36

3. a. If you want 8 more 2-seater canoes than 3-seater canoes, how many of each will you rent?

2-seater canoes = 12

3-seater canoes = 4

b. If you want to have the fewest number of canoes, how many of each will you rent?

2-seater canoes = 0

3-seater canoes = 12

© 2006 ORIGO Education. This page may be reproduced for classroom teaching purposes only.

Algebra for All, Purple Level [Blackline Master | School Trips] 59

Representations 4 — Page 61

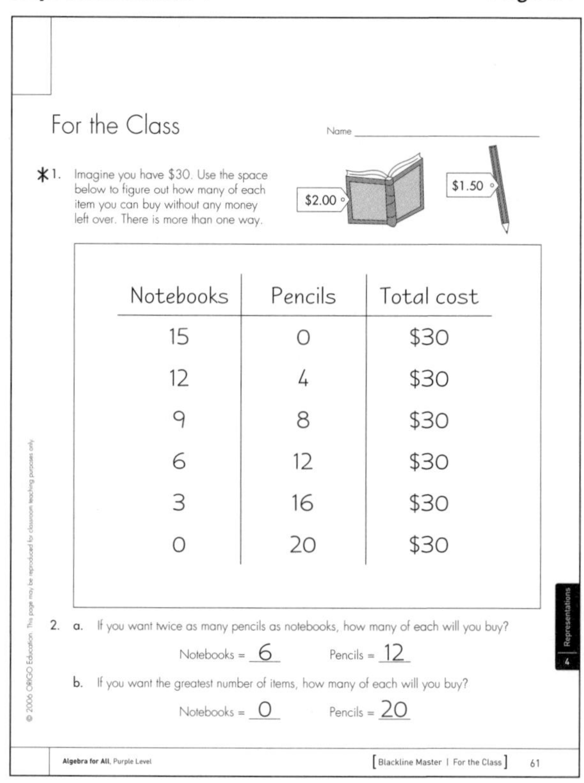

For the Class

Name ____________

*1. Imagine you have $30. Use the space below to figure out how many of each item you can buy without any money left over. There is more than one way.

Notebooks	Pencils	Total cost
15	0	$30
12	4	$30
9	8	$30
6	12	$30
3	16	$30
0	20	$30

2. a. If you want twice as many pencils as notebooks, how many of each will you buy?

Notebooks = 6 Pencils = 12

b. If you want the greatest number of items, how many of each will you buy?

Notebooks = 0 Pencils = 20

© 2006 ORIGO Education. This page may be reproduced for classroom teaching purposes only.

Algebra for All, Purple Level [Blackline Master | For the Class] 61

* Answers will vary. This is one example.

Representations 5 Page 63

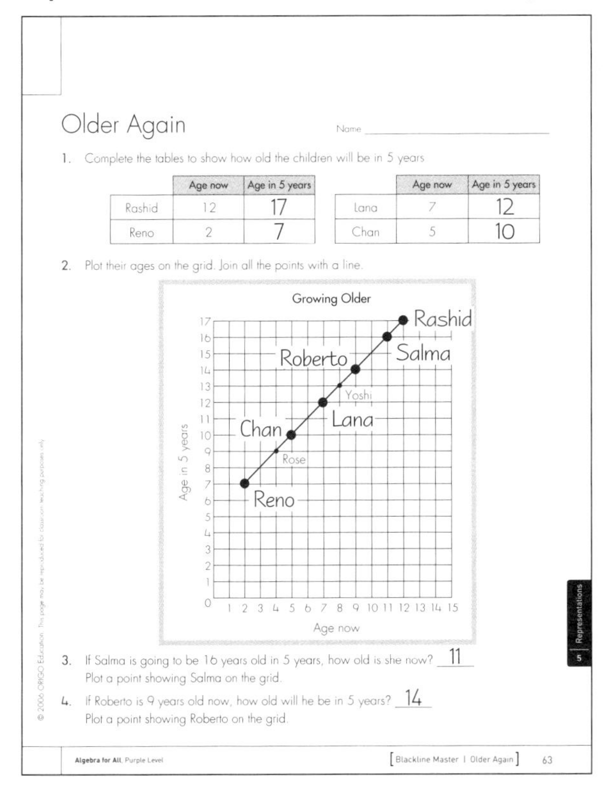

Representations 6 Page 65

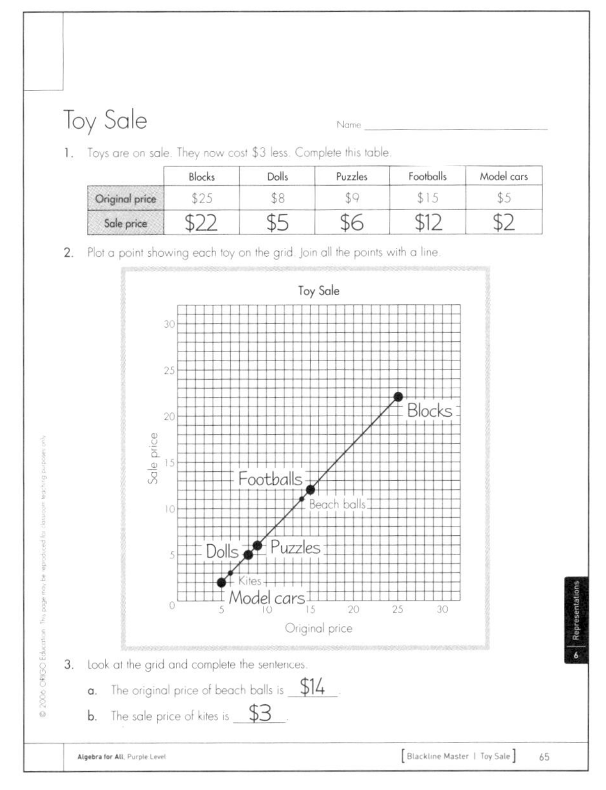

Assessment Summary

Name ______________________________

	Lesson	Page	A	B	C	D	Date
Equivalence and Equations	In the Bag	6					
	Fair Shares	8					
	3D Fun	10					
	Liquid Measures	12					
	Number Stories	14					
	Stepping Stones	16					
	Taller Shorter	18					
	Older Younger	20					
Patterns and Functions	Growing Triangles	22					
	Growing Squares	24					
	Pick the Part	26					
	Pattern Parts	28					
	Making Patterns	30					
	Equal Pieces	32					
	Machine Pairs	34					
	Music Sale	36					
	CDs and DVDs	38					
	Magic Machines	40					
	Double Rules	42					
Properties	Great Groups	44					
	Hot and Cold	46					
	Mix and Match	48					
	Getting Older	50					
	Make it Easy	52					
Representations	Grids Galore	54					
	Home, Sweet Home	56					
	School Trips	58					
	For the Class	60					
	Older Again	62					
	Toy Sale	64					